Principles of
Electroanalytical Methods

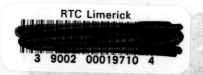

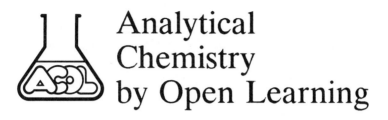

Analytical Chemistry by Open Learning

Titles in Series:

Samples and Standards
Sample Pretreatment
Classical Methods
Measurement, Statistics and Computation
Using Literature
Instrumentation
Chromatographic Separations
Gas Chromatography
High Performance Liquid Chromatography
Electrophoresis
Thin Layer Chromatography
Visible and Ultraviolet Spectroscopy
Fluorescence and Phosphorescence Spectroscopy
Infra Red Spectroscopy
Atomic Absorption and Emission Spectroscopy
Nuclear Magnetic Resonance Spectroscopy
X-Ray Methods
Mass Spectrometry
Scanning Electron Microscopy and Microanalysis
Principles of Electroanalytical Methods
Potentiometry and Ion Selective Electrodes
Polarography and Other Voltammetric Methods
Radiochemical Methods
Clinical Specimens
Diagnostic Enzymology
Quantitative Bioassay
Assessment and Control of Biochemical Methods
Thermal Methods
Microprocessor Applications

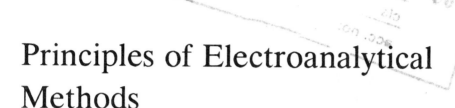

Principles of Electroanalytical Methods

Analytical Chemistry by Open Learning

Authors:
TOM RILEY and COLIN TOMLINSON
Brighton Polytechnic, UK

Editor:
ARTHUR M. JAMES

on behalf of ACOL

Published on behalf of ACOL, London
by
JOHN WILEY & SONS
Chichester · New York · Brisbane · Toronto · Singapore

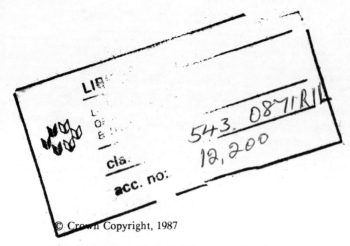

Published by permission of the Controller of
Her Majesty's Stationery Office

Library of Congress Cataloging in Publication Data:

Riley, Tom, 1935–
  Principles of Electroanalytical Methods.
  (Analytical chemistry by open learning)
  Bibliography: p.
  1. Electrochemical analysis—Programmed instruction.
2. Chemistry, Analytic—Programmed instruction.
I. Tomlinson, Colin.   II. James, A. M. (Arthur M.),
1923–   III. ACOL (Firm : London, England)
IV. Title   V. Series.

QD115.R54   1987   543'0871   86–23397

ISBN 0 471 91329 4
ISBN 0 471 91330 8 (pbk.)

British Library Cataloguing in Publication Data:

Riley, Tom
  Principles of Electroanalytical Methods.—
  (Analytical chemistry by open learning).
  1. Electrochemical analysis
  I. Title   II. Tomlinson, Colin   III. James,
  Arthur M.   IV. Analytical Chemistry by
  Open Learning (*Project*)   V. Series

  543'0871   QD115

ISBN 0 471 91329 4
ISBN 0 471 91330 8 Pbk

Printed and bound in Great Britain

Analytical Chemistry

This series of texts is a result of an initiative by the Committee of Heads of Polytechnic Chemistry Departments in the United Kingdom. A project team based at Thames Polytechnic using funds available from the Manpower Services Commission 'Open Tech' Project have organised and managed the development of the material suitable for use by 'Distance Learners'. The contents of the various units have been identified, planned and written almost exclusively by groups of polytechnic staff, who are both expert in the subject area and are currently teaching in analytical chemistry.

The texts are for those interested in the basics of analytical chemistry and instrumental techniques who wish to study in a more flexible way than traditional institute attendance or to augment such attendance. A series of these units may be used by those undertaking courses leading to BTEC (levels IV and V), Royal Society of Chemistry (Certificates of Applied Chemistry) or other qualifications. The level is thus that of Senior Technician.

It is emphasised however that whilst the theoretical aspects of analytical chemistry can be studied in this way there is no substitute for the laboratory to learn the associated practical skills. In the U.K. there are nominated Polytechnics, Colleges and other Institutions who offer tutorial and practical support to achieve the practical objectives identified within each text. It is expected that many institutions worldwide will also provide such support.

The project will continue at Thames Polytechnic to support these 'Open Learning Texts', to continually refresh and update the material and to extend its coverage.

Further information about nominated support centres, the material or open learning techniques may be obtained from the project office at Thames Polytechnic, ACOL, Wellington St., Woolwich, London, SE18 6PF.

How to Use an Open Learning Text

Open learning texts are designed as a convenient and flexible way of studying for people who, for a variety of reasons cannot use conventional education courses. You will learn from this text the principles of one subject in Analytical Chemistry, but only by putting this knowledge into practice, under professional supervision, will you gain a full understanding of the analytical techniques described.

To achieve the full benefit from an open learning text you need to plan your place and time of study.

- Find the most suitable place to study where you can work without disturbance.

- If you have a tutor supervising your study discuss with him, or her, the date by which you should have completed this text.

- Some people study perfectly well in irregular bursts, however most students find that setting aside a certain number of hours each day is the most satisfactory method. It is for you to decide which pattern of study suits you best.

- If you decide to study for several hours at once, take short breaks of five or ten minutes every half hour or so. You will find that this method maintains a higher overall level of concentration.

Before you begin a detailed reading of the text, familiarise yourself with the general layout of the material. Have a look at the course contents list at the front of the book and flip through the pages to get a general impression of the way the subject is dealt with. You will find that there is space on the pages to make comments alongside the text as you study—your own notes for highlighting points that you feel are particularly important. Indicate in the margin the points you would like to discuss further with a tutor or fellow student. When you come to revise, these personal study notes will be very useful.

∏   When you find a paragraph in the text marked with a symbol such as is shown here, this is where you get involved. At this point you are directed to do things: draw graphs, answer questions, perform calculations, etc. Do make an attempt at these activities. If necessary cover the succeeding response with a piece of paper until you are ready to read on. This is an opportunity for you to learn by participating in the subject and although the text continues by discussing your response, there is no better way to learn than by working things out for yourself.

We have introduced self assessment questions (SAQ) at appropriate places in the text. These SAQs provide for you a way of finding out if you understand what you have just been studying. There is space on the page for your answer and for any comments you want to add after reading the author's response. You will find the author's response to each SAQ at the end of the text. Compare what you have written with the response provided and read the discussion and advice.

At intervals in the text you will find a Summary and List of Objectives. The Summary will emphasise the important points covered by the material you have just read and the Objectives will give you a checklist of tasks you should then be able to achieve.

You can revise the Unit, perhaps for a formal examination, by re-reading the Summary and the Objectives, and by working through some of the SAQs. This should quickly alert you to areas of the text that need further study.

At the end of the book you will find for reference lists of commonly used scientific symbols and values, units of measurement and also a periodic table.

Contents

Study Guide

During the past 10 to 15 years electrochemistry has undergone a revival, particularly in its analytical applications. Gone are the days when the main application centred around such techniques as the measurement of pH values or the polarographic analysis of mixtures of metal ions. Not only has the accuracy of these methods been drastically improved but exciting advances have resulted in the establishment of new techniques. This has been helped by advances in instrumentation and electronic circuitry. The development of a range of ion-selective electrodes for both anions and cations, membrane electrodes, biosensors and gas sensing membrane probes has permitted the measurement and continuous monitoring of ions in solution, a wide range of non-ionic organic materials (eg penicillin, urea, glucose) and gases in solution. These techniques have found wide applications in environmental studies and, with miniaturization, in medical fields. New polarographic techniques, some with detection limits of 10^{-8} to 10^{-9} mol dm^{-3}, have been developed and applied to the determination of oxygen in gas mixtures, biological fluids and water in addition to metal ions. Polarographic methods are now widely used in the analysis of organic compounds including naturally occurring mycotoxins (eg fungal metabolites), compounds used to improve food yields (eg insecticides and fungicides) and environmental pollutants (eg nitrosamines, detergents, azo dyes).

Such non-destructive methods, with rapid response times and low detection levels are now in competition with other methods of analysis, such as spectrophotometry and AAS.

This Unit is designed to introduce you to a wide range of electro-analytical techniques. It is intended to impart sufficient knowledge to enable you to understand the basic theory, the practical aspects and the scope of each technique and to enable you to relate the techniques via features they have in common such as electrode and solution behaviour and electrical circuitry. You will not become a practicing expert in any one of these methods—quickly, this will require more study and certainly more practical experience. On completion of this unit you should be able to select a suitable method for a particular application and you will be prepared for further more detailed study of individual techniques.

To achieve these objectives you are first introduced to some of the basic definitions, conventions, theoretical principles and practical approaches of solution electrochemistry. Only those aspects of the subject relevant to subsequent studies of analytical techniques are covered. The remainder of the unit is then devoted to a discussion of about fifteen analytical methods, first those based on electrochemical cells, then those based on electrolysis cells.

It is assumed that you have attained at least level III in Chemistry (BTEC) or GCE 'A'-level with a good understanding of the Physical Chemistry components. In addition, some study of Physics up to 'O'-level or level II (BTEC) would be of benefit in understanding ions, their associated electrical fields, coulombic forces and electrical conduction.

SI units are used extensively, others are used as necesary. Some tables showing the relationship between the various units are provided.

Many texts are quoted as supplementary reading for this unit. Unfortunately, no one book covers all the aspects dealt with here. As a general introduction to the early sections of this unit the monograph by J. Robbins is an excellent, inexpensive text. For an alternative point of view on electro-analytical methods the book by F. W. Fifield and D. Kealey, will serve as a useful accompanying text especially for the later sections.

You are reminded that other Units in the ACOL scheme are devoted to more detailed treatment of selected electro-analytical techniques. Those available at the moment are: Potentiometry and Ion-Selective Electrodes; Polarography and Other Voltammetric Techniques. This unit serves as a good introduction to these two units although there is some overlap of content.

It is hoped that as a result of studying this Unit you will be encouraged to study further in the area of electro-analytical chemistry. If you come to realise the potential and the advantages of some of these techniques and become a user then even better.

Supporting Practical Work

1. GENERAL CONSIDERATIONS

Experimental facilities for electrochemical experiments are generally available in most laboratories. The experiments suggested are designed to occupy a three hour laboratory period using basic equipment; these illustrate the potential applications of electrochemical techniques to analytical problems. If more sophisticated equipment is available and particularly if (as well may be the case) there is need to specialise in a chosen technique, some supplementary experiments are suggested.

2. AIMS

There are three principal aims.

(*a*) To provide basic experience in the choice, preparation, handling and use of electrodes and the correct use of measuring instruments.

(*b*) To illustrate important principles from the theoretical parts of the unit.

(*c*) To illustrate the relevant applications of different electrochemical techniques in analytical chemistry.

3. SUGGESTED EXPERIMENTS

(*a*) Construct some simple galvanic cells using reversible copper, silver and hydrogen electrodes as indicator electrodes and calomel and silver-silver chloride electrodes as reference electrodes. Measure the emf of the cells at different electrolyte concentrations.

(b) Determine the temperature coefficient of a cell (eg Ag, $AgCl/KCl/Hg_2Cl_2,Hg$) and calculate the thermodynamic functions for the cell reaction.

(c) Plot potentiometric titration curves for:

   (i) hydrochloric acid and ethanoic acid against sodium hydroxide using a glass electrode assembly; determine ionization constant of ethanoic acid;

   (ii) mixed halid solution (KCl, KBr, KI) against silver nitrate; determine the concentration of the three halide ions;

   (iii) a redox system, Fe(II) against Ce(IV).

(d) Plot an amperometric titration curve, eg for determination of lead by titration with potassium dichromate solution.

(e) Study polarographic waves produced by oxygen and metals in solution, determine half-wave potentials. Construct a wave-height concentration curve for a given ionic species (eg Cd).

4. SUPPLEMENTARY EXPERIMENTS

(a) Use a potentiostatically controlled three-electrode circuit to investigate the current-potential (I/E) relationship in an electrolysis cell.

(b) Determine the concentration of (i) an analyte in aqeous solution and (ii) an organic analyte in non-aqeous solution by a voltammetric technique after choosing a suitable electrode/supporting electrolyte system.

Bibliography

1. 'STANDARD' ANALYTICAL CHEMISTRY TEXTBOOKS

Virtually all the books giving a comprehensive treatment of (instrumental) analytical chemistry contain one or more chapters on electro-analytical methods of analysis. Appropriate examples are:

(a) F W Fifield and D Kealey, *Principles and Practice of Analytical Chemistry*, International Textbook Co Ltd, 2nd Edn, 1983.

(b) H H Willard, L L Merritt, J A Dean and F A Settle, *Instrumental Methods of Analysis*, Van Nostrand, 1981.

(a) H H Bauer, G D Christian and J E O'Reilly, *Instrumental Analysis*, Allyn and Bacon, 1978.

2. ELECTROCHEMISTRY TEXTBOOKS

(a) D T Sawyer and J L Roberts, *Experimental Electrochemistry for Chemists*, Wiley-Interscience, 1974.

(b) B H Vassos and G W Ewing,, *Electroanalytical Chemistry*, Wiley-Interscience, 1983.

(c) A J Bard and L R Faulkner, *Electrochemical Methods*, Wiley-Interscience, 1980.

(d) J Robbins, *Ions in Solution, 2. An Introduction to Electrochemistry*, Clarendon Press, Oxford, 1972.

(*e*) D R Crow, *Principles and Applications of Electrochemistry*, Chapman and Hall, 1979.

(*f*) A Denaro, *Elementary Electrochemistry*, Butterworths, 1971.

3. MORE SPECIALIZED TEXTS

(*a*) A K Covington, *Ion-Selective Electrode Methodology*, Vols. I, II, CRC Press, 1979.

(*b*) F Franks, *Water*, Royal Society of Chemistry, 1983.

(*c*) D B Hibbert and A M James, *Dictionary of Electrochemistry*, Macmillan, 1984.

NOTES:

Reference 2(*a*) is especially recommended for its treatment of experimental conditions.

Reference 2(*b*) and 2(*c*) present a more advanced treatment of the subject with 2(*c*) being of honours degree standard.

Reference 2(*d*), 2(*e*) and 2(*f*) are standard textbooks which provide a good basic introduction to the more theoretical aspects of electrochemistry.

1. Basic Principles of Solution Chemistry and Electrochemistry

Overview

In this part of the course you are taken through the fundamental ideas on the nature and properties of electrolyte solutions and on electrochemical cells. These ideas will enable you to better understand the methods and practice of electroanalytical chemistry.

Since aqueous solutions have been studied most commonly, these will be used as the vehicle for the discussions. The extension of the ideas to other solvents will be indicated and occasionally developed more fully.

The discussions commence with a consideration of the nature of ions and solvents and then of electrolyte solutions. The non-ideality of these solutions, which arises from electrostatic interactions between the constituent ions, will be stressed. An important and useful aspect of electrolyte solutions is their electrical conduction and this topic is discussed before the electrochemical cell is briefly introduced in anticipation of Part 2.

1.1. IONS IN SOLUTION

Ions are chemical entities containing a multiple of the electronic charge either of a positive or negative nature. In normal chemical situations, we meet ions either as part of a crystal lattice or as components of electrolyte solutions. It is this latter situation which concerns us particularly in electroanalytical chemistry. There can be few simpler solutions than aqueous sodium nitrate. We shall use this solution throughout our discussions on electrolyte solutions and look at it from various points of view. On dissolution of this salt, the sodium and nitrate ions are dislodged from their crystalline lattice positions and dispersed throughout the solvent. For many purposes the solution can be regarded as a single-phase homogeneous system. Thus if we measure the density or refractive index at several places within the solution, the same value would be obtained. These properties are bulk properties of the system. If we begin to look at the solution at the molecular level, we find that it is far from homogeneous. The ions have specific solvent molecules associated with them and the bulk solvent has within it regions which show some intermolecular structure and others without this type of structure.

An ion, an electrically charged entity, will interact electrostatically with any other charged species in the solution. The principal species which we must consider in this regard are other ions and the solvent molecules. Two interactions occur:

(*a*) ion–ion, both between like and unlike–charged ions, and

(*b*) ion–dipole between the electrical dipole in the solvent and any ion.

Other important interactions do occur between the solution entities, eg hydrogen bonding and van der Waals forces; these will be introduced into the discussion as necessary.

1.1.1. Solvent Structure

As a result of X-ray structural investigations on the solid state of solvents it has been established that solvent molecules are to be found on a regular crystal lattice often with precise intermolecular orientations. On the other hand, studies on the vapour state of solvents show that they consist of randomly oriented and positioned molecules. From such investigations the structure of the water molecule has been established (Fig. 1.1a).

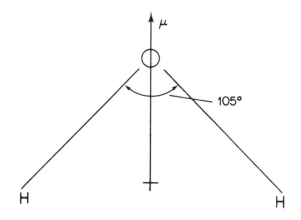

Fig. 1.1a. *The water molecule*

Because of the different electronegativities of the oxygen and hydrogen atoms and because the oxygen atom has two lone pair electron orbitals, the molecule has a permanent electrical dipole moment (μ) indicated by an arrow in the diagram. (Note that, by convention, the arrow head represents the negative end of the dipole). For water the magnitude of this moment is high at 1.87 debye units.

$1 \text{ D} = 3.336 \times 10^{-30}$ C m in SI units. If therefore, we consider an electrical dipole consisting of two electronic charges (1.602×10^{-19} C) separated by 0.1 nm, then the dipole moment (μ) would be:

$$\mu = 1.602 \times 10^{-19} \times 10^{-10} = 1.602 \times 10^{-29} \text{ C m}$$

$$= \frac{1.602 \times 10^{-29}}{3.336 \times 10^{-30}} = 4.80 \text{ D}$$

In terms of molecular dipole moments this is a large value, because the electrical charges are rarely as large as the electronic charge. Some representative dipole moment values are: CO 0.1 D; HCl 1.08 D; HI 0.42 D and CH_3OH 1.71 D.

If we turn now to a consideration of the liquid form of solvents, we should expect their structure to be intermediate between those of the solid and vapour phase. By-and-large this is true. In solvents where the intermolecular forces are strong, the molecules show some mutual ordering effects and the liquid shows evidence of some structure. In solvents where these forces are weak, the molecules in the liquid solvent are randomly orientated as in the vapour state. Solvents exhibiting the former type of behaviour are termed associated and those exhibiting the latter are termed unassociated solvents.

∏   Which of the following common solvents can be regarded as associated?

       Give reasons for your choice.

       water
       tetrachloromethane (carbon tetrachloride)
       propanone (acetone)
       hexane
       methanol

The answer is water and methanol. The molecules of these solvents can hydrogen bond with one another and thus form small regions of structure within the liquid. This leads to unusual liquid properties such as high boiling point and high entropy of vaporisation.

Hexane, being a hydrocarbon, has no significant tendency for hydrogen bond formation. The molecules are held together in the liquid phase by London dispersion forces.

Tetrachloromethane is a similar case, even though there is a carbon–chlorine dipole. The negatively charged chlorine atoms symmetrically arranged on the outside of the molecule repel other like molecules leading to an unstructured solvent.

Propanone is an interesting case for the molecule has a strong carbonyl dipole moment leading to some dipole–dipole binding between propanone molecules. This is however not large enough to lead to unusual properties for this solvent.

As mentioned earlier, water is the most commonly used analytical solvent; let us consider its structure in more detail. Water molecules cannot only interact with each other through the electrostatic dipoles they contain but also through the hydrogen bonds which can form between the oxygen atom of one molecule and the hydrogen atom of another molecule. In ice, a three dimensional lattice is formed where each water molecule is hydrogen bonded to four other water molecules in an open framework structure (Fig. 1.1b).

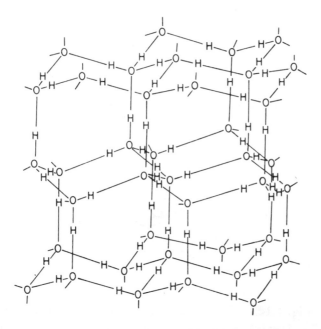

Fig. 1.1b. *The ice structure*

In spite of years of intensive research the exact structure of liquid water still eludes us. Numerous theories have been proposed and we will consider one of these here. Frank and Wen (1957) put forward the flickering cluster model for liquid water. Water is considered to consist of a mixture of free water molecules and water molecules bound-up in ice-like clusters where each molecule is strongly hydrogen bonded to its neighbours. Fig. 1.1c depicts such a mixture. The clusters are thought to have very short lifetimes (approximately 10^{-10}s) before they disintegrate and reform in another part of the liquid. Water is thus seen to be a mixture of ice-like structures and free molecules in a constant state of flux, with any one water molecule being free one moment and bound the next.

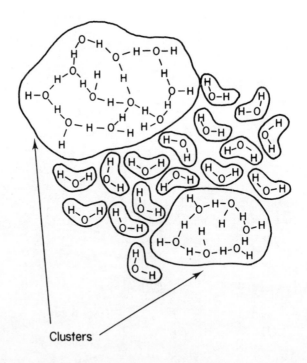

Clusters

Fig. 1.1c.   *The flickering cluster water structure*

An unassociated solvent such as hexane has weak van der Waals forces between its molecules. Since these are similar in all directions the molecules have little desire to have any given mutual orientation and so are mixed up randomly by thermal agitation.

Of course there are a multitude of important solvents between the two extreme types which we have considered above. Those solvents such as propanone or dimethyl sulphoxide have large dipole moments and must therefore interact quite strongly. The liquid state probably contains transient structures involving dimers, chains and rings (Fig. 1.1d).

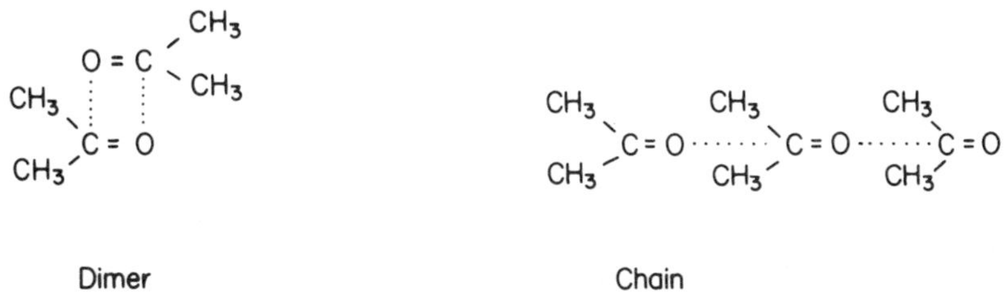

Dimer   Chain

Fig. 1.1d. *Dipolar liquid structures*

In electrochemistry we are concerned with electrolyte solutions and we must now turn to the dissolution of ions in the solvent. If we think of this process then we must first remove the ion from the crystal lattice, create a hole in the solvent to accept it, and finally insert the ion into the hole. Energy expended in the first two of these operations must be recouped in the third. This key process is followed up in the next section.

1.1.2. Ion-solvent Interactions (Solvation)

An ion has an intense electrical field radiating symmetrically outwards from the centre of charge as shown in Fig. 1.1e. The intensity of the field is proportional to the square of the charge and inversely proportional to the distance from the centre of charge. Any solvent molecule containing an electric dipole which experiences this field, will orientate itself so that the dipole lies along the field direction.

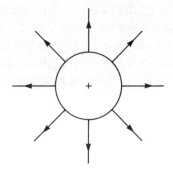

Fig. 1.1e.   *The electric field of an ion*

In this way the energy of the system will be minimised. This tendency must however be modified by the thermal motion of the solvent molecules, which may be sufficient in a weak electric field not to allow any significant alignment of the dipole in the field. Since the field immediately next to an ion is most intense, this is where ion–solvent forces are strongest and alignment occurs. For a cation, the negative end of the solvent dipole will point towards the ion and vice-versa for the anion as depicted in Fig. 1.1f. In this way a new entity has been formed in the solution – the solvated ion or in the case of water, the hydrated ion.

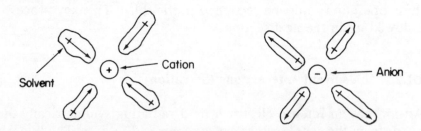

Fig. 1.1f.   *Dipolar solvation of ions*

If we consider the case of water, the first hydration shell usually contains either four or six molecules, eg $Al(H_2O)_6^{3+}$. The presence of such structures has been demonstrated experimentally in a few favourable cases. Consider the proton magnetic resonance spectrum shown in Fig. 1.1g.

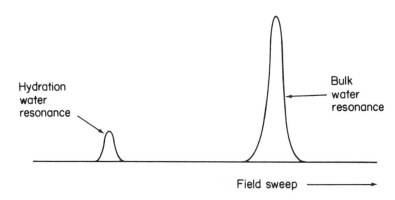

Fig. 1.1g *Proton magnetic resonance spectrum, Al(NO$_3$)$_3$ in water*

You will notice that there are two proton resonance peaks for the water molecules in the system. That at low field comes from the hydration water, because the cation has drawn electrons away from the protons, so deshielding these nuclei. The peak at higher fields arises principally from the bulk water molecules.

Having dealt with the region immediately next to the ion, what can we say about that further away? It is easy to propose that the solvent structure a long way away from the ion will differ little from bulk solvent structure. In between these two extremes therefore there must be a region of solvent structure where the molecules are being influenced as much by the field of the ion as by the intermolecular forces of the solvent. The solvent molecules in this region are consequently oriented randomly.

Our discussion above has brought us to a neat physical picture of ionic solvation which is summarised in Fig. 1.1h. Solvent molecules are categorised as:

(*a*) *inner or primary* solvent. These solvation molecules are interacting strongly with the ion and thus have different properties to those of the normal solvent molecules;

(*b*) *outer or secondary* solvent. These molecules are not aligned either by the field of the ion or by the forces from the bulk solvent. This region around the ion is one of broken structure and of high entropy;

(*c*) *bulk* solvent. This outermost region contains solvent molecules as in the normal solvent.

∏   The alkali metal ions lithium, sodium and potassium have crystal ionic radii of 0.06, 0.095, and 0.133 nm respectively. How would you expect the heat of hydration to differ between these ions? Consider only the primary hydration sphere.

The electrical field intensity near to an ion is inversely proportional to the distance from its centre. This field must therefore be most intense near the lithium ion. Since the interaction between an ion and a water dipole depends on the size of this field, the lithium ion must have the highest hydration energy, followed by sodium and then potassium.

The literature values for the enthalpy of hydration are, lithium -129.7, sodium -102.3, potassium -82.3 kJ mol^{-1}.

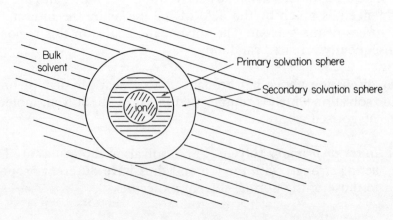

Fig. 1.1h.   *Overall view of ionic solvation*

This picture of solvation can be applied to all solvents but the relative importance of each region will differ in each case. Let us now return to the aqueous solution with which we started, that of sodium nitrate, and look at the effects which arise through one ion interacting with another, the ion–ion interaction.

SAQ 1.1a

> Consider the first layer of solvent molecules around an ion. Which of the items below do you think are important in determining the number of solvent molecules in this layer?
>
> (*i*)   ionic size;
> (*ii*)   ionic charge;
> (*iii*)   solvent structure;
> (*iv*)   solvent molecular volume.

SAQ 1.1b   Classify the following solvents as associated or
                otherwise:
                benzene; ethanol; trichloromethane.

SAQ 1.1c   Explain with reasons which of the following
                molecules you would expect to have a zero
                dipole moment:
                H_2; H_2O; Ar; NH_3; CCl_4; $CHCl_3$; CO; CO_2;
                C_6H_6; $C_6H_5NO_2$.

SUMMARY AND OBJECTIVES

Summary

A liquid solvent has a structure intermediate between that of its solid and gas state structures. Associated solvents, eg water, have a structure resembling that of the solid state and unassociated solvents, eg hexane, have a structure akin to the gas state. Ionic solvation derives from the electrostatic force between the ion and surrounding solvent molecules. In water, hydration is viewed in terms of two solvation spheres, the primary and secondary, which surround the ion. The water in these solvation shells differs from bulk water in its properties.

Objectives

You should now be able to:

● explain the structure of liquid solvents;

● give an account of the nature of ionic solvation.

1.2. ION–ION INTERACTIONS

In a very very dilute solution (usually curiously termed an infinitely dilute solution), the ions must naturally be so far apart that the influence of their electric fields upon one another must be negligibly small. This is an ideal ionic solution which we will consider in more detail later. For the moment we wish to study the effects as the solution concentration is increased and the ionic separation decreases.

The electrostatic force between two ions follows Coulomb's Law which for our purpose we will write as:

$$\text{Coulombic force} \propto \frac{z_+ z_- e^2}{\epsilon_r r^2} \qquad (1.2a)$$

where, z_+ and z_- are the ionic charge numbers (eg 1 for sodium
    ion, -2 for sulphate ion),
  e the charge on the electron (1.602×10^{-19} C),
  ϵ_r the relative permittivity of the solvent between the ions,
    and
  r the distance between the ions.

This Coulombic force thus follows an inverse square law with respect to ionic separation, increasing rapidly as this separation dimin-
ishes. The force is repulsive for ions of like charges and attractive
for ions of opposite charges. An important feature of Coulombic
forces is their inverse dependence on the relative permittivity of the
solvent (also known as the solvent dielectric constant). Values of this
parameter for some common solvents are listed in Fig. 1.2a. You will
notice that ϵ_r for water is higher than for many other common solvents, so minimising ion–ion interactions in this solvent. You should
memorize the above equation and appreciate the effects of charge,
distance and relative permittivity upon inter-ionic forces.

Solvent	ϵ_r
N-methyl formamide (NMF)	190.5
methanoic acid (formic acid)	111.5
water	78.5
dimethyl sulphoxide (DMSO)	46.4
ethanonitrile (acetonitrile)	37.5
methanol	34.0
ethanol	23.8
propanone (acetone)	20.7
dichloromethane	8.9
tetrachloromethane	2.2

Fig. 1.2a.   *Relative permittivity of various solvents at 298 K*

Consider now the situation around a sodium ion in our sodium nitrate solution. Other sodium ions will be repelled from its vicinity
but nitrate ions will be attracted. Thermal motion of the ions will

be attempting to mix them randomly. On average however, there will be more nitrate ions near the sodium ion than other sodium ions. We describe this situation by saying that the cation has an atmosphere of negative charge surrounding it. This is depicted in Fig. 1.2b. Because the solution has no net charge, the total negative charge in the atmosphere must equal the charge on the central cation.

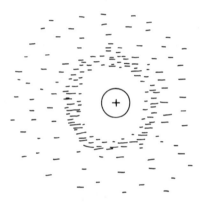

Fig. 1.2b. *The ionic atmosphere*

This concept of the ionic atmosphere arises from the theory of Debye and Hückel (1923) in which they derived an equation for the effective radius of the atmosphere:

$$\text{radius of atmosphere} \propto \left(\frac{\epsilon_r T}{I}\right)^{\frac{1}{2}} \qquad (1.2b)$$

where, T/K is the temperature, and I the ionic strength.

The ionic strength of an electrolyte solution measures the total electrostatic-interaction capability of that solution. It can be quite simply calculated for completely dissociated electrolytes from:

$$I = 0.5\sum c_i z_i^2 \qquad (1.2c)$$

where, c_i is the concentration of ion i in the solution,
z_i the charge number of the ion i, and $\sum$ is the sum of $c\,z^2$
terms for each ion in the system.

∏   What will be the ionic strength of an 0.1 mol dm^{-3} aqueous
solution of calcium chloride?

Before we start the calculation an assumption must be made re-
garding the degree of dissociation of the electrolyte. In this case it
is reasonable to assume that it is fully dissociated so that the ionic
concentrations are:

$$c(Ca^{2+})\ (mol\ dm^{-3})\ =\ 0.1$$
$$c(Cl^{-})\ (mol\ dm^{-3})\ =\ 0.2$$

(Remember that calcium chloride dissociates into one calcium ion
but two chloride ions per mole.)

Thus, $I = 0.5(0.1 \times 2^2 + 0.2 \times 1^2) = 0.3$ mol dm^{-3}

If we return to a consideration of Eq. 1.2b, we see that the atmosphere radius becomes smaller as ϵ_r decreases. Again therefore we
note that ion–ion interactions are more important in general in solvents other than water. Note also that because ionic charge increases
I, the radius of the atmosphere decreases as ionic charge increases
because of the enhanced ion–ion effects.

The ionic atmosphere is a very important aspect of electrolyte solutions. Because of it, the properties of the central ion are modified.
If you consider the movement of the central ion, the electrostatic
force exerted by the atmosphere will resist this movement. In other
ways too the properties are altered, some of these changes will be
discussed in later sections.

Before coming to this however, let us consider the consequences
when the Coulombic forces become very large, so large in fact that
the atmosphere closes in to the central ion and the cation and anion
are almost in contact. From our discussions above, this situation
is most likely to arise in solvents of low relative permittivity and

with ions of high charge. The two ions now form a new entity –
an ion pair. There are many possible structures for such a species
but two are illustrated in Fig. 1.2c. This new entity will have quite
different properties from the individual ions for it has less charge
(may be zero if the two ions have the same charge) than these ions
and therefore more solubility in non-polar solvents.

Contact

Solvent separated

Fig. 1.2c. *Ion pair structures*

If we pursue the properties of this new entity, we can write a dissociation equilibrium:

$$(Na^+, NO_3^-) \overset{K}{\rightleftharpoons} Na^+ + NO_3^-$$

ion pair   'free' ions

The equilibrium constant, K, has been measured for numerous systems, a selection of values is listed in Fig. 1.2d. You will notice that
the values of K are surprisingly small even for some common electrolytes and that the presence of ion pairs in electrolyte solutions
must always be considered as a possibility.

Ion pair	K
Na^+, OH^-	5
Na^+, NO_3^-	4
Li^+, OH^-	1.2
Ca^{2+}, OH^-	0.04

Fig. 1.2d. *Ion pair dissociation constants in water at 298 K*

(Ion pairing is a fascinating subject and becoming of increasing application in analytical chemistry, eg high performance liquid chromatography. See the bibliography for further reading matter.)

1.2.1. Ionic Activity

When dealing with gases, an ideal gas is defined as one in which the intermolecular forces are zero. The gas obeys the ideal gas equation:

$$pV = RT$$

Real gases do not follow this equation at normal pressures because of significant intermolecular forces, but they approach the ideal gas behaviour at very low pressures.

In the liquid phase, we can no longer think in terms of zero forces between molecules since the very existence of the liquid depends on such forces. Our concept of an ideal solution must therefore be a system in which the intermolecular forces remain constant as the solute concentration changes. In such a system the Gibbs energy (free energy) of the solute is given by,

$$G = G^{\ominus} + RT \ln c \qquad (1.2d)$$

where, G is the Gibbs energy (free energy) of the solute, $G^{\ominus}$ the standard Gibbs energy of the solute, c the solute concentration.

The equation would apply to ions if they formed an ideal solution. We have seen however, that the ion–ion forces are dependent on the ionic concentration and these solutions must be non-ideal. The ion–ion interaction brings into consideration another energy term – the Coulombic interaction term – which must be added into Eq. 1.2d:

Coulombic
interaction term

$$G = G^{\ominus} + RT \ln c + RT \ln y$$
$$= G^{\ominus} + RT \ln a \qquad (1.2e)$$

where y is the solute activity coefficient and
a the solute activity.

Comparison of Eq. 1.2d and 1.2e shows that activity must replace concentration in thermodynamic equations, eg equilibrium constant, electrochemical cell equations. What is activity and how is it related to solute concentration?

Activity is the true measure of the behaviour of a solute in solution but is, unfortunately, an experimental quantity. It is linked to solute concentration by the solute activity coefficient, y.

$$a = yc \tag{1.2f}$$

Once the activity has been measured, the value of y can readily be deduced.

(You should be aware that there are analogous equations to Eq. 1.2f, involving molal concentration units ($a = \gamma m$) and mole fraction units ($a = fx$). The activity coefficients y, γ and f are numerically almost identical at low concentrations (< 0.001 mol dm^{-3}).)

It is not possible to measure the activities of cations and anions separately in an electrolyte solution. Consequently we must define mean molar ionic quantities for the activity ($a_{\pm}$), activity coefficient ($y_{\pm}$), and concentration ($c_{\pm}$) of an electrolyte. These parameters are linked by an analogous equation to 1.2f,

$$a_{\pm} = y_{\pm} c_{\pm} \tag{1.2g}$$

For an electrolyte which contains p positive ions and n negative ions per mole, it can be shown that for an electrolyte concentration, c, the corresponding mean molar ionic concentration is given by,

$$c_{\pm} = c(p^p n^n)^{1/(n + p)} \tag{1.2h}$$

∏   Calculate the mean molar concentration for 0.1 mol dm^{-3} calcium chloride solution.

We can write down the quantities in terms of the symbols of Eq. 1.2h. Thus, $c = 0.1$ mol dm^{-3}, $p = 1$, $n = 2$.

Hence $c_{\pm} = 0.1 \times (1^1 \times 2^2)^{1/3}$

                   $c_{\pm} = 0.159 \text{ mol dm}^{-3}$

Mean molar ionic activity coefficients are listed in data books for various electrolyte concentrations in water at 25 °C. For other solvents and temperatures, data is not always readily obtained.

1.2.2. Ionic Activity Coefficient

If you look at Fig. 1.2e you will see how the activity coefficient varies with solute concentration. Note that the diagram uses the convention for describing electrolytes: an M : N electrolyte is one dissociating into a cation of charge M and an anion of charge N.

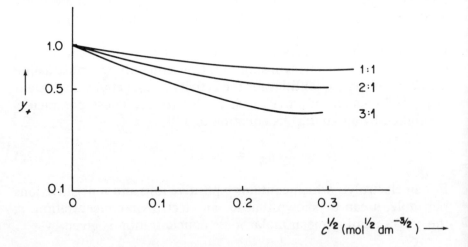

Fig. 1.2e.   *Mean ionic molar activity coefficients in water at 25 °C*

For all electrolytes the value of $y_{\pm}$ approaches unity as the concentration approaches zero. This must be true since the ion–ion interactions are becoming very small at low concentrations – the solution is approaching ideality. As the concentration is increased the activity coefficient falls below unity, ie the ionic activity is less than the ionic concentration. This effect is more marked where ions of higher

Open Learning

21

charge type are involved. At very high ionic concentrations, $y_\pm$ for some electrolytes rises steeply and may have values much in excess of unity. This effect is due to ionic solvation, the solvent removed from the solution in this way increasing the effective concentration of the solute. It was mentioned earlier that $y_\pm$ is an experimental quantity. However, it is possible to calculate its value in very dilute solution (<0.001 mol dm^{-3}) using the Debye–Hückel Limiting Law:

$$-\lg y_\pm = A \,|\, z_+ z_- \,|\, I^{1/2} \qquad (1.2\text{i})$$

where, A is the Debye–Hückel constant for each solvent at a given temperature, (Fig. 1.2f), and $|\, z_+ \, z_- \,|$ is the positive value of the ionic charge product.

solvent	$A/\text{mol}^{-1/2} \text{ dm}^{3/2}$
water	0.511
ethanonitrile (acetonitrile)	1.41
methanol	1.69
propanone	3.33

Fig. 1.2f. *Values of the Debye–Hückel constants (A) at 298 K for various solvents*

Π   Study Fig. 1.2e and decide whether the following statements are true or false.

(*i*) $y_\pm$ for 0.09 mol dm^{-3} aqueous calcium chloride solution at 25 °C is 0.5.

(*ii*) at low concentrations the value of $y_\pm$ decreases linearly as the square root of the concentration increases.

(*i*) TRUE. If $c = 0.09$, then $c^{1/2} = 0.3$. Calcium chloride is a $2:1$ electrolyte since calcium ion has a charge of 2 and chloride ion a charge of one. From the figure, on the $2:1$ line at $c^{1/2} = 0.3$ the $y_\pm$ value is 0.5.

(*ii*) FALSE. Although a casual glance at the figure might indicate
that there is the linear relationship between $y_\pm$ and c, a closer
look shows that the $y_\pm$ axis is a logarithmic scale. Thus the
relationship is linearity between lg $y_\pm$ and $c^{1/2}$ Eq. 1.2i.

This theoretical equation was developed with the assumptions that
ions were point charges and were not solvated. These assumptions
lead to the calculated $y_\pm$ values at higher concentrations being lower
than the measured values. Many empirical equations have been devised to calculate $y_\pm$ at these higher concentrations but none are
totally satisfactory.

Π   Estimate the mean molar ionic activity coefficient in 0.001
         mol dm^{-3} aqueous potassium chloride solution at 25 °C.

Your first task is to estimate the ionic strength of the solution in a
similar way to that used previously.

It must be assumed that the electrolyte is fully dissociated and therefore the ionic concentrations are:

$$c\,(K^+)\,(\text{mol dm}^{-3}) = 0.001$$
$$c\,(Cl^-)\,(\text{mol dm}^{-3}) = 0.001$$

The ionic strength follows using Eq. 1.2c.

$$I = 0.5\,(0.001 \times 1^2 + 0.001 \times 1^2) = 0.001 \text{ mol dm}^{-3}$$

This value can now be fed into Eq. 1.2i using the value for A given
above:

$$-lg\,y_\pm = 0.511 \times 1 \times 1 \times (0.001)^{\frac{1}{2}}$$
$$= 0.162$$
$$y_\pm = 0.963$$

This is a sensible value for $y_\pm$ since it is a dilute solution of a 1 : 1
electrolyte and a value just below unity is expected (see Fig. 1.2e).
It demonstrates that the ions will be about 4% less effective in equilibria and electrochemical situations than their concentration would
suggest.

SAQ 1.2a

> You are considering changing the solvent in an electrochemical experiment from water to a series of methanol/water mixtures. If your electrolyte behaves as a strong electrolyte in water judge whether the following factors are important or not in making this solvent change.
>
> (*i*) solute activity;   YES / NO
> (*ii*) electrolyte solubility;   YES / NO
> (*iii*) chemical nature of the electrolyte ions;
>   YES / NO
> (*iv*) ionic association.   YES / NO

SAQ 1.2b

> Estimate the mean molar ionic activity for the ions of lanthanum nitrate, $La(NO_3)_3$, in water at 25 °C in a 0.005 mol dm^{-3} solution.

SAQ 1.2c   Calculate the ionic strength of a mixture of 20 cm^3 of 0.10 mol dm^{-3} KNO$_3$ and 40 cm^3 of 0.20 mol dm^{-3} CaCl$_2$ which has been diluted to a final volume of 100 cm^3.

SAQ 1.2d   Calculate the ionic strength of a mixture of 20 cm^3 of 0.1 mol dm^{-3} sodium ethanoate and 10 cm^3 of 0.1 mol dm^{-3} hydrochloric acid made up to a total volume of 50 cm^3.

SUMMARY AND OBJECTIVES

Summary

Ions in solution have a mutual electrostatic interaction leading to the concepts of ionic atmosphere and solution non-ideality as expressed through ionic activity. This empirical quantity can be approached theoretically in dilute ionic solutions by the calculation of the ionic activity coefficient using the Debye–Hückel Limiting Law.

Objectives

You should now be able to:

- discuss the implications of the Coulomb equation for ion–ion interactions in solution.

- explain the theoretical basis of activity and the factors which govern the value of the ionic activity coefficient.

1.3. IONIC MIGRATION IN ELECTROLYTE SOLUTIONS

The migration of an ion in an electrolyte solution can be caused either by an electrical potential difference or a concentration gradient across the solution. The ion therefore undergoes electrical ionic migration or concentration diffusion. The two effects are related through the parameter ionic mobility.

1.3.1. Electrical Ionic Migration

Electrical current originates in the movement of electrical charge. Current is defined as the charge passing a given point in the circuit per unit time. If a charge of one coulomb (C) passes per second, a current of one ampere (A) is said to flow. By convention, the current is considered to flow in the direction of the positive charge movement.

In metals the current is carried by electrons. This mechanism is occasionally true in solution but for electrolyte solutions the current is carried mainly by the movement of the anions and cations. Under the influence of an applied potential, the cations and anions will move in opposite directions in the solution. The total current flowing will be the sum of two contributions:

$$I(\text{total}) \;=\; I_+(\text{anion}) \;+\; I_-(\text{cation}) \qquad (1.3a)$$

All types of electrical conduction obey Ohm's Law. Thus if a potential difference (V) is applied across a solution of resistance (R) the current flowing is:

$$I \;=\; V/R \qquad (1.3b)$$

For electrolyte solutions the inverse of the solution resistance is termed the *conductance* (G) and is measured in units of ohm^{-1} or siemens (S), thus:

$$I \;=\; V\,G \qquad (1.3c)$$

Solution conductance can therefore be considered as the current which will flow through the solution upon application of a potential difference of one volt.

∏   A saturated solution of barium sulphate in water is found to have a resistance of 0.5 MΩ at 25 °C when this is measured in a conductance cell. In the same cell, water has a resistance of 5.0 MΩ. Calculate the conductance of the barium sulphate in the solution.

We must calculate the conductance of the solution and the water since these can then be subtracted to give the electrolyte conductance.

$$
\begin{aligned}
G(\text{solution}) &= 1/R = 1/(0.5 \times 10^{6}) \\
&= 2.0 \times 10^{-6}\ \text{S} \\
G(\text{water}) &= 1/(5.0 \times 10^{6}) \\
&= 0.2 \times 10^{-6}\ \text{S}
\end{aligned}
$$

$$G(BaSO_4) = G(\text{solution}) - G(\text{water})$$
$$= (2.0 - 0.2) \times 10^{-6}$$
$$= 1.8 \times 10^{-6} \text{ S}$$

Cation	$10^8\ u^v$/m^2 V^{-1} s^{-1}	Anion	$10^8\ u^v$/m^2 V^{-1} s^{-1}
H^+	36.24	OH^-	20.58
Li^+	4.01	F^-	5.74
Na^+	5.19	Cl^-	7.92
K^+	7.62	Br^-	8.09
Ag^+	6.42	NO_3^-	7.41
Mg^{2+}	5.50	ClO_4^-	6.98
Cu^{2+}	5.56	SO_4^{2-}	8.25
Zn^{2+}	5.50		
La^{3+}	7.21		

Fig. 1.3a. *Limiting ionic mobilities in water at 298 K*

Conductance, like current, is an additive property for the solution. Each ionic species in the solution contributes an amount of conductance to the total. The amount which any particular ion-type contributes depends upon its ionic velocity and charge as well as on its concentration.

Ionic velocity is usually discussed in terms of *ionic mobility* (u). This is defined as the velocity of an ion under a unit potential gradient. If we use the units of m s^{-1} for velocity and V m^{-1} for potential gradient, the units of mobility are m^2 V^{-1} s^{-1}. Some typical values of limiting ionic mobility (ie in infinite dilute solution) are listed in Fig. 1.3a

In order to standardise the results of conductance measurements, the conductance of a one metre cube of solution which contains one mole of a given ion is designated the *ionic molar conductivity* (λ). The units of this parameter are S m^2 mol^{-1}. Because of ion–ion electrostatic interactions, λ varies with ionic strength. Tabulated λ-values are therefore at infinite dilution, ie λ^∞. The value of λ^∞ can be obtained from the corresponding limiting ionic mobility,

$$\lambda^\infty = u^\infty z F \qquad (1.3d)$$

∏   Using the data in Fig. 1.3a, calculate the limiting ionic molar conductivity of the sodium and magnesium ions in water at 25 °C. Given $F = 96485 \, Cmol^{-1}$.

Your answer should be 50.08×10^{-4} S mol^{-1} m^2 for the sodium ion and 106.1×10^{-4} S mol^{-1} m^2 for the magnesium ion. Consider the sodium ion, $z = 1$ so that from Eq. 1.3d,

$$\lambda^\infty = u^\infty(Na^+) \times 1 \times F$$
$$= 5.19 \times 10^{-8} \times 96485 \text{ m}^2 \text{ V}^{-1} \text{ s}^{-1} \text{ C mol}^{-1}$$
$$\therefore \qquad \lambda^\infty (Na^+) = 50.08 \times 10^{-4} \text{ S m}^2 \text{ mol}^{-1}$$

Similarly for the magnesium ion but setting $z = 2$ we have,

$$\lambda^\infty(Mg^{2+}) = 5.50 \times 10^{-8} \times 2 \times 96485$$
$$\therefore \qquad \lambda^\infty(Mg^{2+}) = 106.1 \times 10^{-4} \text{ S m}^2 \text{ mol}^{-1}$$

It is interesting to look at Fig. 1.3a in more detail. You will see that the lithium ion, because of its larger hydration shell, has a lower mobility than the potassium ion. Similar arguments can be applied to the fluoride and bromide ions. Exceptional mobilities are observed for the hydrogen and hydroxide ions. These result from a special conduction mechanism for these ions. Look at Fig. 1.3b which shows the hydrogen ion (H_3O^+) in water. Notice that hydrogen bonds have been drawn-in between the ion and the oxygen of the neighbouring water molecule and from this to its neighbour, etc. If the hydrogen atom *a* moves within its hydrogen bond to form a bond with oxygen atom 2 rather than with atom 1, the hydrogen ion has effectively moved a distance to the right which is large by comparison with the actual hydrogen atom movement. This conduction mechanism is more akin to charge than ion movement. It depends for its success on the peculiar structure of water and is therefore only found for solvent-derived ions in hydrogen-bonded solvents, eg NH_4^+ in liquid ammonia.

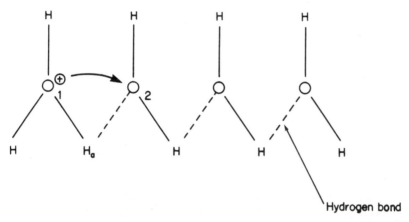

Fig. 1.3b. *The hydrogen ion conduction mechanism*

The fraction of the total current which an ion carries through a solution is known as the *transport number* (*t*). Thus,

$$t_+ = \frac{I_+}{I}$$

$$t_- = \frac{I_-}{I}$$

where, t_+, t_- are the cation and anion transport numbers.

In general t must lie between zero and unity, be high for ions of high mobility and low for ions of low mobility. Fig. 1.3c shows these features clearly. The electrolyte KCl is interesting from an electrochemical point of view for the two ions carry almost equal proportions of the current, this fact is made use of in salt bridges (Part 2).

Electrolyte	Conc/mol dm^{-3}	t_+
HCl	0	0.821
	0.01	0.825
	0.10	0.831
	3.0	0.843
LiCl	0	0.336
NaCl	0	0.396
KCl	0	0.491
AgNO$_3$	0	0.464

Fig. 1.3c.   *Ionic transport numbers in water at 298 K*

(The values quoted for zero concentration have been obtained by extrapolation to infinite dilution from finite concentrations).

∏   It has been described above how a moving charge constitutes electrical current. Using this idea, calculate the transport number for the hydrogen ion in a dilute aqueous solution of hydrochloric acid from the mobility values given in Fig. 1.3a.

The listed mobilities are 36.24×10^{-8} and 7.92×10^{-8} m^2 V^{-1} s^{-1} for the hydrogen and chloride ions, respectively. Since both ions carry a single electrical charge, these mobility values must be proportional to the currents carried by these two ions.

Current carried by hydrogen ion ∝ 36.24

Current carried by chloride ion ∝ 7.92

∴   total current ∝ 36.24 + 7.92 = 44.16

∴   fraction of current carried by the hydrogen ion is
$\dfrac{36.24}{44.16} = 0.821$

This ratio is in fact the transport number of the hydrogen ion and tallies with value quoted in Fig. 1.3c.

1.3.2. Ionic Concentration Diffusion

We have seen from Eq. 1.2d that the free energy of an ion in solution increases with its concentration. If therefore, a concentration gradient exists in a solution, the electrolyte will diffuse from the more concentrated part to the less concentrated part of the solution. Given time, the concentration throughout the solution will equalise itself. Concentration gradients are often a key part of electroanalytical systems, eg polarography, and their detailed behaviour will be discussed in later Parts.

The mass diffusion rate of ions due to a concentration gradient is proportional to both the size of the concentration gradient and the ionic diffusion coefficient (D). The diffusion coefficient bears a simple relationship to the ionic mobility,

$$D = k\,T\,u/z \qquad (1.3e)$$

where, k is the Boltzmann constant.

This completes our introduction to ionic processes in solution which are of relevance to electroanalytical chemistry. Now we turn our attention to the electrode/electrolyte solution interface and the chemical and physical phenomena of interest there.

SAQ 1.3a

In an electrochemical experiment, an aqueous electrolyte solution containing 0.1 mol dm^{-3} potassium chloride and 0.001 mol dm^{-3} zinc sulphate is the analyte solution. What percentage of the electrical current which passes through this solution will be carried by the zinc, potassium and chloride ions?

(Make use of the data in Fig. 1.3a. Neglect the effect of non-ideality on ionic mobility).

SAQ 1.3a

SUMMARY AND OBJECTIVES

Summary

Cations and anions carry the electrical current through electrolyte solutions. The ability of any particular ion to carry the current depends upon its charge and its mobility. The ionic mobility is related to the ionic conductance and to the ionic diffusion due to a concentration gradient.

Objectives

You should now be able to:

- understand the nature of electrical mobility and ionic concentration diffusion.

- define the terms conductance, molar conductivity and transport number of an ion and understand the relationships between these three quantities.

1.4. AN INTRODUCTION TO ELECTRODES AND THE ELECTROCHEMICAL CELL

In order to pass an electrical current through an electrolyte solution, we must introduce two electrical conductors (*electrodes*). An electrode can be as simple as a wire, eg copper, silver, or as complex as some ion-selective electrodes. (A full discussion of the construction and properties of electrodes will be found in Part 2).

The system consisting of two electrodes dipping into an electrolyte solution is known as an *electrochemical cell* (Fig. 1.4a) or an *electrolysis cell.* Such cells are well known to us in the form of batteries and electroplating cells. Note that an electrochemical cell itself develops a potential difference between the electrodes as is the case with a battery, but in an electrolysis cell a potential difference is imposed on the cell from an external source as is the case in electroplating. In both instances, a chemical reaction occurs within the cell. Each type of cell finds application in electroanalytical chemistry. We will look at the chemical reactions occuring at an electrode and the consequences of this when equilibrium prevails at the electrode and also when a significant current is allowed to pass through the electrode.

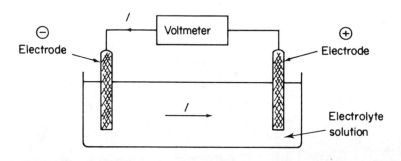

1.4.1. The electrode

A convenient type of electrode is a clean piece of metal dipping into a solution of its own ions, eg a copper wire in an aqueous copper(II)sulphate solution. Two reactions now become possible. Firstly, the metal can throw off its ions into the solution, leaving the electrons behind on the metal:

$$M(s) \longrightarrow M^{z+}(aq) + ze$$

Secondly, metal ions from the solution can combine with electrons at the metal surface to form metal atoms:

$$M^{z+}(aq) + ze \longrightarrow M(s)$$

These two reactions are complementary and form a dynamic equilibrium at the electrode surface:

$$M(s) \rightleftharpoons M^{z+}(aq) + ze$$

If the forward reaction predominates over the reverse reaction, the metal will become negatively charged with respect to the solution and vice-versa (Fig. 1.4b). In this manner a potential difference is established across the electrode/solution interface.

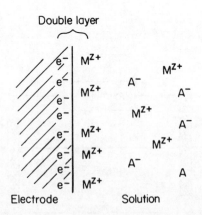

Fig. 1.4b.   *The electrode solution interface*

This potential difference is termed the *electrode potential* (E). Its value depends not only on the nature of the metal but also on the activity of the metal ions in the solution. The relationship between the electrode and the metal ion activity is known as the *electrode response.*

The steady state at the electrode/solution interface can be modified by allowing a current to flow across the interface. This occurs in many practical applications of cells, eg electrodeposition. In this new situation an electrode reaction occurs appropriate to the needs of the current passing. Thus, if current is being drawn from the electrode through the external circuit, then electrons must be produced by the forward electrode reaction which must then occur to a much greater extent.

∏   Write down the reaction which occurs at the electrode/solution interface for a silver electrode in silver nitrate solution. If electrons are removed from this electrode by passing a current into it, what chemical consequence will result at the electrode?

The electrode reaction is:

$$Ag(s) \rightleftharpoons Ag^+(aq) + e$$

If electrons are removed from the silver by the external circuit then these must be provided by the electrode reaction, ie the forward reaction must predominate. Silver must therefore dissolve and pass into the solution to provide the necessary electrons.

1.4.2. The Electrode/Solution Interface

Let us now return to a consideration of the state of affairs in the electrolyte solution near to an electrode which has a negative potential upon it. The electrical field stretches out into the solution from the electrode as shown in Fig. 1.4c.

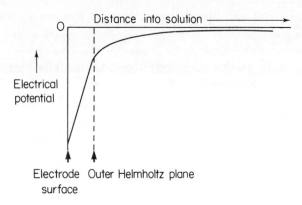

Fig. 1.4c.   *The electrical potential at the electrode/solution
interface*

The electrical field falls linearly with distance through the
Helmholtz–Perrin region. In this region there are basically two types
of ion. Firstly there are contact ions absorbed directly on the electrode surface. These are unsolvated ions. This is known as the *inner
Helmholtz plane* (Fig. 1.4d). Secondly, there are adsorbed solvated
ions, the *outer Helmholtz plane*.

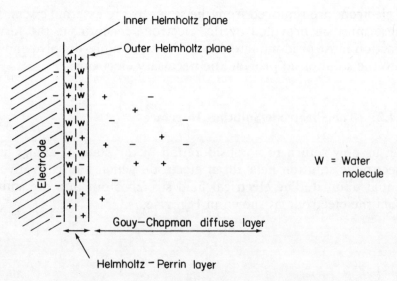

Fig. 1.4d.   *The electrode/solution interface*

You will see a direct analogy here to contact and solvated ion-pairs in an electrolyte solution. The Helmholtz–Perrin region often contains a large percentage of the ionic charge necessary to balance the charge on the electrode and thus the electrical potential is considerably less at the outer Helmholtz plane than at the electrode surface. The ions in the solution outside this plane are under two opposing forces: electrical forces and thermal forces; consequently a diffuse region – the Gouy–Chapman diffuse layer is formed. This region is similar to the ionic atmosphere which forms around any ion in solution.

Omitted so far from the argument is any mention of the role of water molecules. These are attracted to the electrode surface because of their electrical dipole and form a layer there together with the contact adsorbed ion. These water molecules are in a quite different situation to those in bulk water and have different properties, eg a relative permittivity of 6 as against 78.5.

∏   In a polarographic experiment the potential of a mercury electrode in potassium chloride solution is changed from a positive to a negative potential. What changes should occur at the electrode/solution interface?

When the electrode potential is positive, chloride ions will adsorb at the electrode surface and the oxygen atom of the surface water molecules will point at the surface. When the electrode changes to a negative potential the potassium ions will adsorb at the electrode surface and the hydrogen atoms in the surface water molecules will point at the surface.

1.4.3. Faraday's Laws

In many electroanalytical techniques, eg coulometry, a significant electrical current is passed through the solution. This has the consequence of producing significant amounts of chemical changes at the electrodes. Faraday (1833) formulated two laws concerning this which, in view of our discussions above, are fairly self-evident. Let us consider these laws in terms of the two electrode reactions which occur at the silver and copper electrodes,

$$Ag^+ + e \longrightarrow Ag$$

$$Cu^{2+} + 2e \longrightarrow Cu$$

One electron or one faraday of charge can liberate one mole of silver. In the case of copper, one faraday only liberates one half mole of copper. These facts encapsulate the Faraday laws.

Law 1

The amount of a substance liberated or dissolved at an electrode is proportional to the amount of electricity passed.

Law 2

The amounts of different substances liberated or dissolved at different electrodes for the same quantity of electricity passed, are proportional to their molar masses divided by the number of electrons in the respective electrode processes.

It is clearly important to be able to measure the quantity of electricity (Q) passed in any experiment. This is related to the current (I) flowing and the time (t),

$$Q = I t \tag{1.4a}$$

For a steady current, there is no problem assessing Q, but for a varying current an integral of the current–time plot must be made. A common device for achieving this is a chemical coulometer in which the amount of a chemical deposited or liberated in the time of the experiment is determined gravimetrically or volumetrically and by applying Faradays laws, Q can be calculated.

It should be noted that the value of the faraday is 96485 C mol^{-1}. This is the total electrical charge of one mole of singly charged cations or anions.

Π A current of 3 mA is passed for 100 minutes through an electrolysis cell containing a silver electrode dipping into silver nitrate solution. Calculate the mass of silver deposited on the silver electrode ($A_r(Ag) = 107.9$).

You must first calculate the quantity of electricity passed in the experiment by using Eq. 1.4a,

time $= 100$ min $= 6000$ s

current $= 3$ mA $= 3 \times 10^{-3}$ A

$Q = I \times t = 3 \times 10^{-3} \times 6000 = 18$ C

But we know that one mole of silver will be deposited for one faraday of electricity passed.

Thus, for 18 C of electricity, 18/96485 mole of silver are deposited. Since the relative atomic mass of silver is 107.9: mass of silver deposited $= (18/96485) \times 107.9 = 0.02013$ g

SAQ 1.4a A cell is constructed from two copper wires dipping into aqueous copper(II)sulphate solution. What effects will be observed if:

(*i*) the two copper wires are joined together outside the cell,

(*ii*) a low voltage battery is connected between the two copper wires?

SAQ 1.4b

> An electric current was passed through a series of solutions of $AgNO_3$, $CrCl_3$, $ZnSO_4$ and $CuSO_4$. 1.000 g of silver was deposited from the first solution. Calculate (*i*) the quantity of electricity passed (*ii*) the current passing, and (*iii*) the weights of Cr, Zn and Cu deposited simultaneously [Ag = 107.9; Cr = 52.0; Zn = 65.4; Cu = 63.6].

SUMMARY AND OBJECTIVES

Summary

A cell consists of two electrodes placed in an electrolyte solution. At the electrode/solution interface a steady state potential difference develops which is characteristic of the electrode and the concentration of the ion in solution to which the electrode is responsive. If the steady state is modified by passing a current across the interface either electrodeposition or dissolution occur according to Faraday's laws.

Objectives

You should now be able to:

● write down the reaction(s) which occur at an electrode surface;

● explain the structure at an electrode solution interface;

● use Faraday's Laws.

2. Galvanic Cells

Overview

In Part 2 we study electrochemical cells, ie cells which generate
an emf spontaneously by converting the free energy of a chemical reaction directly into electrical energy. These cells contain two
electrodes separated by an electrolyte. The potential developed at
any electrode depends upon the activities of substances within the
electrode and the electrolyte solution. This electrode response can
therefore be utilised for the measurement of these activities. A wide
variety of electrodes exists: the key types are discussed with respect
to their response, electrode chemistry and general structure. You
will learn how to formulate the electrode reactions and response
and how to write down the cell in a conventionally accepted manner.

The measurement of cell emf and pH are given special attention
but the implications of the theory to electroanalytical chemistry is
stressed throughout. The final Section draws together many of the
ideas discussed to explain a potentiometric titration in detail.

2.1. GALVANIC CELLS

It is convenient to identify two general types of cell. Firstly, the *Galvanic cell* (or *electrochemical cell*) is one which develops its own emf spontaneously due to a chemical reaction occuring within it. Batteries produced commercially are galvanic cells. Such a cell can drive a current through an external circuit. In electroanalytical chemistry, the galvanic cell is commonly used to monitor the concentration of an analyte. In this case the current flowing is very near to zero and the cell is operating close to equilibrium or thermodynamic conditions. We will discuss this application, called potentiometry (Section 2.8).

The second type of cell is the *electrolytic cell* (or *electrolysis cell*) where a current is forced through it from an external power source. The effects of the current flow are monitored for analytical purposes, eg in coulometry the mass of a substance electrolytically deposited is a measure of the electrical charge passed.

Let us return to a consideration of the galvanic cell. There is a direct connection between the cell emf, E/V and the Gibbs (free) energy change, $\Delta G/J$ of the spontaneous cell reaction (Eq. 2.1a)

$$\Delta G = -zEF \qquad (2.1a)$$

If we apply this equation to a cell having an emf of 1 V then, if $z = 1$, the free energy change is -96.49 kJ. The free energy change of chemical reactions is commonly 0 to -200 kJ and we therefore expect emf values of electrochemical cells to lie in the range 0 to 2 V.

To set the scene for our discussions on electrochemical cells, let us briefly look at an interesting example – the Bacon fuel cell – which was used in the NASA space programme (Fig. 2.1a).

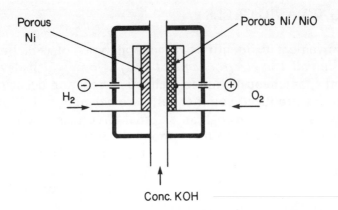

Fig. 2.1a. *Bacon fuel cell*

In this cell hydrogen is oxidised and oxygen is reduced at nickel-based electrodes, in a concentrated potassium hydroxide electrolyte. The chemical reaction occuring in the cell (for $z = 2$) is,

$$H_2 \text{ (g)} + 0.5\, O_2 \text{ (g)} \longrightarrow H_2O \text{ (l)}$$

This cell generates an emf of approximately 1.2 V and precious water for the astronauts as a reaction product! The free energy change for the cell reaction is therefore $-2 \times 1.2 \times 96.5 = -232$ kJ per mole of water. If you consult thermodynamic data tables you can confirm that this is a sensible answer (look up the standard free energy of formation of liquid water).

∏   A galvanic cell has the reaction:

$$0.5\, H_2(g) + 0.5\, Cl_2(g) \longrightarrow HCl(g)$$

This corresponds to a one electron exchange at the electrodes. We have another cell whose reaction we can write as:

$$H_2(g) + Cl_2(g) = 2\, HCl(g)$$

How are the standard free energies of the two cell reactions related and are the two cell emf's the same?

The answer is that the second reaction has twice the standard free energy change of the first reaction and the two cells have the same emf.

The free energy change in a reaction is the sum of free energy of the products less the sum of the free energy of the reactants. Since the second reaction refers to twice the molar quantities of the first reaction, the free energy change must be double that of the first reaction.

Because there are double molar quantities involved in the second reaction, the number of electrons in the electrode processes must also be double. From Eq. 2.1a, a doubling of ΔG and a doubling of z leaves the value of E unchanged. It does not matter therefore, in galvanic cell work, whether we use multiples of a cell or electrode reaction when calculating the corresponding emf.

One further point is worth noting about the above electrochemical cell: there are two electrodes joined by an electrolyte solution. Each electrode and its associated electrolyte is known as a *half cell*. In general the electrolyte is not common to the two electrodes and the cell must then contain a liquid–liquid junction. This can be achieved by separating the two electrolyte solutions with a porous ceramic or glass frit.

SAQ 2.1a   Predict the emf of a galvanic cell which has the following cell reaction ($z = 1$),

$$Ag\ (s)\ +\ 0.5\,Cl_2(g)\ \longrightarrow\ AgCl(s)$$

$$\Delta G^{\ominus} = -109.8\ kJ$$

Assume that the chlorine gas pressure is 1 atmosphere.

SAQ 2.1a

SUMMARY AND OBJECTIVES

Summary

A galvanic cell derives its emf from the free energy change of the spontaneous chemical reaction occuring in the cell: $\Delta G = -zEF$.

Objectives

You should now be able to:

● explain the relationship between the emf of a galvanic cell and the free energy change of the reaction occuring in the cell.

2.2. ELECTRODE TYPES

In electroanalytical chemistry the emf of a galvanic cell is monitored to indicate the progress of the analysis. The potential of one electrode, the *reference* electrode, is assumed to have a constant value, $E(\text{ref})$. The calomel electrode is one such electrode.

The potential of the other electrode changes as the concentration of the analyte or related species changes. This electrode is consequently called the *indicator* electrode with a potential, $E(\text{ind})$. In the measurement of pH using a pH meter the indicator electrode is the glass electrode. For the analytical cell we may now write the emf of the cell:

$$E = E(\text{ind}) - E(\text{ref}) \qquad (2.2a)$$

Because there is such a large variety of electrode types we will only discuss a few of them here. Ion-selective electrodes in particular will be left for subsequent discussion.

All electrodes need much care in preparation and use. Practical details of these two aspects will occasionally be mentioned but fuller details will be found in the literature.

2.2.1. Metal/Metal Ion Electrodes

These electrodes are often referred to as First Order or Class I electrodes. They consist of the metal dipping into a solution of its ions. The electrode potential depends on the activity of the metal and the metal ion and the electrode is thus an indicator of the metal ion activity, if the metal activity is constant.

A silver rod dipping into silver nitrate solution or a cadmium-plated platinum wire dipping into cadmium ion solution would constitute electrodes of this type. At the electrode surface reduction (ie gain of electrons) would occur:

$$Ag^+ + e \quad \rightleftharpoons Ag$$

$$Cd^{2+} + 2e \quad \rightleftharpoons Cd$$

(in general $M^{z+} + ze \rightleftharpoons M$)

In the first case the passage of one faraday of electrical charge would correspond to one mole of silver being deposited.

In the second case, one mole of cadmium would require two faradays of charge for its liberation. In their use in electrochemical cells however, the current flowing would be close to zero and there would be no significant chemical change occurring.

Electrodes of the above type can be formed in several ways, usually by electrochemical deposition of the metal on an inert base, eg platinum wire, or by chemically or mechanically creating a clean metal surface. Unfortunately, this convenient and simple form of electrode can only be used for a few metals, notably Ag, Cu, Cd, Pb and Hg.

Metal/metal ion electrodes give an electrical response to the metal ion activity in solution. The response is said to be Nernstian, ie it follows Eq. 2.2b:

$$E(M^{z+},M) = E^{\ominus}(M^{z+},M) + (RT/zF)\ln a(M^{z+}) \quad (2.2b)$$

where, $E(M^{z+},M)$ is the electrode potential and $E^{\ominus}(M^{z+},M)$ the standard electrode potential (ie the potential when $a(M^{z+}) = 1$.

Many metals do not give a reliable potential in a solution of their ions because of strains in the metal structure or because of an oxide surface coating. Included among such metals are Fe, Ni, Co, W, Al. Two electrode types which may be considered as derivatives of the metal/metal ion type are the gas and amalgam electrodes.

2.2.2. The Gas Electrode

In this electrode the metal is replaced by a gas and this is in equilibrium with a solution of its ions. The gas/electrolyte equilibrium must occur on the surface of an inert metal or other conductor so that electron provision and removal can be achieved.

Let us see how these features are catered for in the case of the hydrogen electrode (Fig. 2.2a). Pure hydrogen is split into a stream of fine bubbles by the glass frit and passes upwards over the surface of the platinum black. Here the hydrogen is adsorbed and participates in the oxidation/reduction reaction with hydrogen ions from solution.

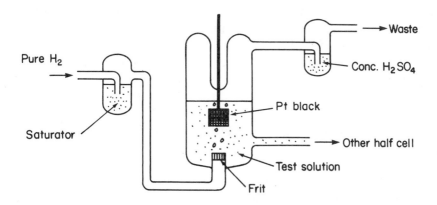

Fig. 2.2a. *Hydrogen electrode*

$$H^+ + e \rightleftharpoons 0.5\,H_2$$

Note that this reaction is analogous to that for the metal/metal ion electrode.

The excess hydrogen gas passes out through the bubbler and escapes to atmosphere. The bubbler prevents the ingression of oxygen which would react with the surface hydrogen atoms and upset the proper working of the electrode. The response of this electrode is:

$$E(H^+, H_2) = E^{\ominus}(H^+, H_2)$$
$$+ (RT/F)\ln\{a(H^+)/p^{1/2}(H_2)\} \qquad (2.2c)$$

where, $a(H^+)$ is the hydrogen ion activity in solution, and

$p(H_2)$ the partial pressure of hydrogen gas.

This is a very important electrode since it is the ultimate standard for electrochemical electrode potentials; all other electrode potentials are referred to it. It must be prepared with great care, the hydrogen gas must be pure and the solutions must not contain substances which would poison the platinum black catalytic surface, eg arsenic, mercury or sulphur, or oxidising agents.

The standard hydrogen electrode (SHE) is constructed with this hydrogen ion and hydrogen gas at unit activity. For a gas, activity is more usually termed fugacity. In practical terms these quantities are achieved with a solution concentration of hydrogen chloride near 1.2 mol dm^{-3} and a gas partial pressure near to 1 atmosphere. From the above equation we see that under these conditions the electrode potential is given by $E^{\ominus}(H^+,H_2)$. We shall soon see that this quantity is arbitrarily set at zero for all temperatures.

Other gas electrodes can be prepared in a similar manner to the hydrogen electrode.

∏   A chlorine gas electrode is constructed by passing chlorine over a platinum black surface in an aqueous chloride solution. By analogy with the hydrogen electrode, write down the electrode reaction (NB this must be a reduction reaction) and its response.

Your electrode reaction should be,

$$0.5\ Cl_2 + e \rightleftharpoons Cl^-$$

Also acceptable would be any multiple of this reaction.

When looking at the reaction for an electrode you must find a chemical species common to the electrode and the electrolyte. In this case this is chlorine atom/ion. These two species are linked together by the electron and must be written into the equation so that this is the case. We see this above, the chlorine atom ($0.5\ Cl_2$) upon addition of the electron gives the chloride ion.

The electrode response is:

$$E(Cl_2,Cl^-) = E^{\ominus}(Cl_2,Cl^-) + (RT/F)\ \ln\ \{p^{1/2}(Cl_2)/a(Cl^-)\}$$

If you did not get this answer, look again at Eq. 2.2b and Eq. 2.2c. The format of these two equations is the same with the logarithmic term containing the activities of the species in the electrode equation. (You should note that pure solids have unit activity and are omitted from the equation). The ratio is always (reactants/products), the activity of each species being raised to the power indicated by the molar quantity in the electrode reaction. Thus the pressure of chlorine is raised to the power of $\frac{1}{2}$.

2.2.3. The Amalgam Electrode

Consider for a moment, how we could construct a sodium class I electrode, so that we could monitor the concentration of sodium ions in solution. If we tried to use pure sodium, there would be a very rapid reaction indeed! However, by lowering the sodium activity we can reduce this reaction rate to manageable proportions so that a reproducible electrode is formed. The lower metal activity is achieved by amalgamating the sodium at low concentration, eg 5% wt/wt. This type of electrode has an electrode reaction,

$$M^{z+} + ze \rightleftharpoons M(Hg)$$

The electrode is unusual in being liquid.

The electrical response equation must contain a term for the metal activity as well as the normal term for the metal ion activity,

$$E(M^{z+}, M) = E^{\ominus}(M^{z+}, M)$$
$$+ (RT/zF) \ln \{a(M^{z+})/a(M)\} \qquad (2.2d)$$

2.2.4. Anion Electrodes

These electrodes are often referred to as Second Order or Class II electrodes. Among electrodes in this class are some of the most widely used and important electrodes, the calomel and silver/silver chloride reference electrodes. The electrical response of these electrodes depends upon the anion activity and they are therefore indicator electrodes for this parameter. To form an electrode responsive to a particular anion, a sparingly soluble salt of this anion must be found. Thus for the chloride ion, we can think in terms of silver chloride or mercury(I) chloride (calomel). For the sulphate anion, mercury(I) sulphate is suitable. The electrode is then constructed as shown in Fig. 2.2b.

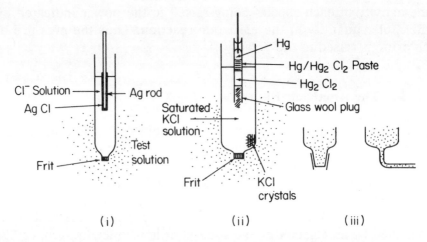

Fig. 2.2b.   *Reference electrodes (i) silver/silver chloride (ii) calomel*
(iii) alternative leak constructions

Let us consider the two most important members of this class of
electrode.

2.2.5.   The Silver/Silver Chloride Electrode

The electrode reaction is effectively:

$$AgCl(s) + e \rightleftharpoons Ag(s) + Cl^-$$

and the electrical response is:

$$\begin{aligned} E(AgCl,Ag) &= E^{\ominus}(AgCl,Ag) \\ &+ (RT/F) \ln \{1/a(Cl^-)\} \end{aligned} \quad (2.2e)$$

This is widely used as a reference electrode as a half cell or as a
component of another electrode, eg the glass electrode. It can be
easily constructed by electrolysing a chloride solution with a silver
anode, so forming an electrolytic layer of silver chloride over the
silver base. This electrode will then yield a potential according to
Eq. 2.2e when dipped into a chloride ion solution. On the other
hand, if it is placed in a solution of fixed chloride ion activity it

will act as a fixed potential electrode, ie a reference electrode. For a 1.0 mol dm^{-3} potassium chloride solution at 25 °C the potential is 0.222 V. The temperature coefficient is -0.5 mV K^{-1}. An attractive property of this electrode is its ability to operate at temperatures up to 275 °C and in non-aqueous solutions. It can also be made very compact.

2.2.6. The Calomel Electrode

In all of electrochemistry this must be the most widely used electrode. Fig. 2.2b (ii) shows a typical format for the electrode.

The electrode reaction is:

$$Hg_2Cl_2 + 2e \rightleftharpoons 2\,Hg + 2\,Cl^-$$

and the electrical response is:

$$E(\text{cal}) = E^\ominus(\text{cal}) + (RT/F)\ln\{1/a(Cl^-)\} \qquad (2.2f)$$

As with the silver/silver chloride electrode, the potential of the calomel electrode depends only on the chloride ion concentration, and a reference electrode is readily made by immersing the electrode in a fixed-concentration chloride solution.

$c(KCl)$/mol dm^{-3}	$E(\text{cal})$/V
0.1	$0.334 - 8.8 \times 10^{-5}(T\text{-}298)$
1.0	$0.280 - 2.8 \times 10^{-4}(T\text{-}298)$
Saturated	$0.241 - 6.6 \times 10^{-4}(T\text{-}298)$

Fig. 2.2c. *Electrode potentials of calomel electrodes of different KCl concentrations*

Fig. 2.2c lists the common forms of the calomel reference electrode and the corresponding electrode potentials. You should note the very low temperature coefficient of the 0.1 mol dm^{-3} electrode.

This electrode does have some disadvantages. It cannot be used at temperatures in excess of 80 °C because of the disproportionation of the mercury(I) ion into mercury and the mercury(II) ion. If the electrode is subject to temperature changes, it equilibrates quite slowly.

An important consideration for both the silver/silver chloride and the calomel electrode reference half cells is the outflow of potassium chloride solution into the analyte solution. This, for certain of the leakage constructions, eg the sleeve, may be as low as 10^{-3} cm^3 h^{-1}. However, since saturated KCl solution is approximately 3.5 mol dm^{-3}, this can affect the analysis either because of chloride ion contamination or through ionic strength changes.

2.2.7.   The Redox Electrode

This electrode type is sometimes known as the Unattackable Electrode type. As the name implies, the electrode is inert and usually made from a foil or wire of a noble metal, eg Pt, Au. This dips into the electrolyte solution which contains both the oxidised and the reduced forms of the chemical species. The metal therefore merely acts as an electron conductor in much the same way as in the case of the gas electrode except that there the metal also had a catalytic role. As an example, let us consider a shiny, clean piece of platinum wire dipping into an aqueous solution of iron(II) and iron(III) ions. At the metal surface the following redox reaction can occur:

$$Fe^{3+} + e \rightleftharpoons Fe^{2+}$$

giving an electrode electrical response of:

$$E(Fe^{3+}, Fe^{2+}) = E^{\ominus}(Fe^{3+}, Fe^{2+})$$
$$+ (RT/F) \, ln \, \{a(Fe^{3+})/a(Fe^{2+})\} \quad (2.2g)$$

The potential is dependent upon the ratio of the activities of the two iron ions and can thus be used to monitor this ratio. Some redox reactions occur in conjunction with acid conditions and the electrode response reflects this. Thus for the redox couple, $Cr_2O_7^{2-}$, Cr^{3+} the electrode reaction is,

$$Cr_2O_7^{2-} + 14\,H_3O^+ + 6\,e \rightleftharpoons 2\,Cr^{3+} + 21\,H_2O$$

The electrode response can be written down as,

$$
\begin{aligned}
E(Cr_2O_7^{2-},\ Cr^{3+}) &= E^{\ominus}(Cr_2O_7^{2-},\ Cr^{3+}) \\
&\quad + (RT/6F)\ln\{a^{14}(H^+)a(Cr_2O_7^{2-})/a^2(Cr^{3+})\}
\end{aligned}
$$

2.2.8. Membrane Electrodes

Electrodes of this type have been developed more than any other in recent times because of their many applications in analytical situations, particularly where an electrical transducer is required to continuously monitor an analyte, eg process quality control, medical work, safety and environmental monitoring. In this regard, you should note that an electrical transducer is ideal for coupling, through an interface, to a computer data logging/analysis system. In addition, electrodes can be made very compact, are cheap, and do not usually interfere with analyte solutions.

For any one type of membrane electrode, the electroactive membrane separates the analyte solution from an internal reference electrode (Fig. 2.2d). The membrane chosen must respond to the analyte ions being monitored, eg the glass membrane for the hydrogen ion, the lanthanum fluoride crystal membrane for the fluoride ion.

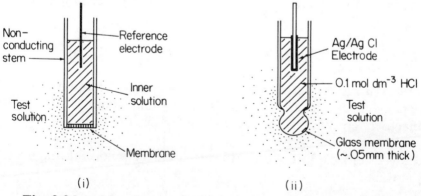

(i)   (ii)

Fig. 2.2d. *Membrane electrodes (i) General construction (ii) Glass electrode*

Thus if we look at Fig. 2.2d, we see that for the glass electrode, a thin glass membrane separates the analyte solution, which contains hydrogen ions, from the inner reference electrode which is a silver/silver chloride electrode in a solution of HCl. At the membrane surface, an electrical potential is developed which is dependent upon the activity of the analyte ions being measured. The inner membrane surface interfaces with a solution of a fixed activity of the ions to which it is responsive and thus has a fixed potential, E(inner). Consequently, the overall response of the electrode is made up as follows:

$$E(\text{memb}) = E(\text{inner}) + E(\text{ref}) + E(\text{outer})$$
$$E(\text{memb}) = E'(\text{memb}) + (RT/F) \ln (a_i) \qquad (2.2h)$$

where, (a_i) is the activity of the analyte ions, and E'(memb) the sum of all the potentials of the electrode which are constant.

Any particular membrane does not give an electrical response exclusively to one ion but often responds significantly to a range of ions. It is said to be *ion selective* in its response towards one type of ion. In the case of the glass electrode, it can be made selective towards the hydrogen ion by proper choice of the glass composition. Nevertheless, it still responds to some extent to other ions, notably the sodium ion.

To express the preference of an electrode for a given ion, a more extensive equation than (2.2h) is used, this is the Nicolsky–Eisenman equation 2.2i,

$$E(\text{memb}) = E'(\text{memb}) + (RT/F) \ln \{(a_i) + K_{ij}(a_j)^{zi/zj}\} \quad (2.2i)$$

where, (a_j) is the activity of any other ion to which the electrode is responsive, z_i and z_j the charges on the ions i and j, and K_{ij} the selectivity constant of the membrane for ion i over ion j.

Let us briefly explore the significance of this equation through a test question.

∏ You are told that a glass electrode which you wish to use in pH measurements has a selectivity constant of 10^{-10} for the hydrogen ion over the sodium ion. Estimate the error in your measurement of pH of a 10^{-2} mol dm^{-3} sodium hydroxide solution. Assume that the activity coefficients are unity.

Your answer should be 0.30 pH units.

From Eq. (2.2i) you can write the logarithmic term for our case here as,

$$\ln \{a(H^+) + K_{ij}\, a(Na^+)\}$$

We are told that: $K = 10^{-10}$; $a(H^+) = 10^{-12}$; $a(Na^+) = 10^{-2}$.

The logarithmic term thus becomes:

$$\ln(10^{-12} + 10^{-10} \times 10^{-2}) = \ln (2 \times 10^{-12})$$

But pH $= -\lg a(H^+) = \lg 2 \times 10^{-12} = 11.7$

For the 10^{-2} mol dm^{-3} solution of NaOH the pH should be $-\lg 10^{-12} = 12.0$.

The error in pH reading is consequently 0.30 units.

Numerous membranes have been developed which are responsive to both anions and cations. Ion selective electrodes are classified by the membrane type. Thus we have *solid state membrane, liquid membrane, glass membrane* and *gas sensing ion-selective* electrodes. In addition, there are the recent *enzyme electrodes* which are based upon ion-selective electrodes but which are responsive to a particular substrate within the solution being analysed.

Our main consideration here is with the glass electrode for pH measurement, but before coming to this let us summarise the general nature of the other types of ion-selective electrodes mentioned above.

Solid state membrane electrodes have a membrane constructed from an insoluble salt either as a crystal or as a matrix with silicone rubber or other inert material. The salt is the active medium and the electrode is responsive to one of the ions in the insoluble salt. In this way electrodes with a response to F^-, CN^-, Cl^-, Br^-, I^-, Ag^+, Pb^{2+}, Cu^{2+} and Cd^{2+} can be produced.

Liquid membrane electrodes are of two types. In both types the membrane consists of a PVC-gel incorporating the active ion-selective substance. In one type this material is a liquid ion-exchanger and in the other it a ligand substance which selectively binds a given ion. Electrodes responsive to Ca^{2+}, K^+, NO_3^-, ClO_4^- etc have been constructed.

Gas sensing electrodes are employed for monitoring dissolved or liberated gases in solution, eg SO_2, CO_2, NO_2, NH_3. The gas diffuses through a polymer gas-permeable membrane into an inner liquid whose pH is measured with a glass electrode/reference electrode combination. Now we shall return to the glass electrode and look at its construction and operation in some detail.

2.2.9. The Glass Electrode

At the turn of this century, it was discovered that a glass/electrolyte interface developed a potential dependent upon the pH of the electrolyte solution. This eventually led to the successful production of the glass electrode for pH measurement. We have already looked at the basic structure of the electrode but a glance at Fig. 2.2d (*ii*) will remind you of this. The inner electrode solution is 0.1 mol dm^{-3} HCl. This fixes the potential of the inner glass surface as well as that of the silver/silver chloride electrode.

Glass has a structure of a framework of SiO_4 tetrahedral units joined into a three dimensional lattice with charge-balancing cations in some of the interstices. A typical chemical composition (wt/wt) for an early glass was 72% SiO_2, 22% Na_2O and 6% CaO. This is a soft soda glass which is easily worked into the thin membrane required.

The resistivity of glass is very high, consequently the membrane has a resistance of several megohms. A high input-impedance voltmeter (eg a pH meter) must be used with the electrode. As the temperature decreases the glass resistance increases exponentially. This makes it difficult to operate glass electrodes at low temperatures.

Prior to use, the glass electrode should be conditioned by soaking in a 0.1 mol dm^{-3} HCl solution (or other acidic buffer solution) for several hours. This procedure hydrates the outer layer (0.1 μm) of the glass surface forming a gel coat which is capable of exchanging its charge-balancing cation with a cation in the surrounding electrolyte solution:

$$Na^+ \quad + \quad H^+ \quad \rightleftharpoons \quad Na^+ \quad + \quad H^+$$
$$\text{(glass)} \quad \text{(electrolyte)} \quad \text{(solution)} \quad \text{(glass)}$$

The extent to which this reaction proceeds depends upon the glass and the solution pH. For a given glass, the electrical response is:

$$E\text{(glass)} \ = \ E'\text{(glass)} \ + \ (RT/F) \ln a(H^+)$$
$$E\text{(glass)} \ = \ E'\text{(glass)} \ - \ (2.303 \, RT/F) \, \text{pH} \qquad (2.2j)$$

This theoretical response is drawn out in Fig. 2.2e together with the typical response of a glass electrode.

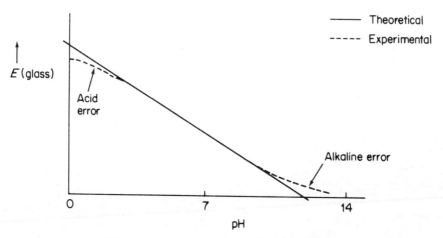

Fig. 2.2e. *Glass electrode response*

You should notice that there are two parts of the pH range where the response of an electrode is likely to be poor. The *alkaline error* occurs at high pH and is due to the electrode responding to alkali metal cations. The *acid error* occurs at low pH and is caused by a lack of water to form an adequate gel-layer on the glass surface. Manufacturers of glass electrodes usually supply a nomograph which allows a correction to be made for these errors.

SAQ 2.2a

You have been given the task of devising an electrochemical cell for the estimation of chloride in a commercial product which has an aqueous base. Draw up a short list of half cells which you consider suitable for this purpose and then make a choice of two electrodes giving your reasons based on the criteria of:

(*i*)   response;
(*ii*) practicability;
(*iii*) availability;
(*iv*) safety.

SUMMARY AND OBJECTIVES

Summary

Electroanalytical galvanic cells are usually constructed of an indicator and a reference electrode. The response of the indicator electrode takes the general form,

$$E = E^{\ominus} + 2.303\,(RT/F)\,\lg\,(\text{activity term})$$

Electrodes can be classified by their general structure and response into four classes:

Class I are electrodes responsive to metal ions, in particular. Gas electrodes such as the hydrogen electrode, belong to this class. Class II electrodes are responsive to anions and include the important examples of the silver/silver chloride and the calomel reference electrodes. Redox electrodes are responsive to a redox couple in the cell electrolyte. Membrane electrodes can be made responsive to many anions and cations with the most widely used example as the glass electrode.

Objectives

You should now be able to:

● define the terms indicator and reference electrode;

● describe the structure and electrical response of Class I, Class II, redox and membrane electrodes.

2.3. IUPAC CONVENTIONS

Electrochemistry has been notorious for the confusion among scientists concerning the meaning of the term electrode potential and the sign convention associated with such potentials. In 1953, a set of conventions was adopted by the International Union of Pure and

Applied Chemistry (IUPAC) and these are accepted today by the majority of electrochemists. The conventions are listed below in a form convenient for our usage. Beware that there are still many textbooks in circulation which do not adhere in part or total to these conventions.

When we study a galvanic cell experimentally, we can establish two facts namely the value of the cell emf (E) and the electrode with the higher potential. This electrode carries the higher positive charge and thus is the positive electrode.

Convention 1 The cell is written down with its positive electrode as the right hand electrode.

A direct consequence of this convention is that if we join the two electrodes together through, say, a voltmeter (Fig. 2.3a), then electrical current flows spontaneously through the external circuit from right to left, and through the cell itself from left to right.

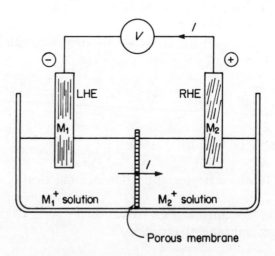

Fig. 2.3a.   *Electrochemical cell*

Electron flow must be opposite to the current flow. Because electrons flow to the right hand electrode (RHE) in the external circuit, these must be removed by the reaction at this electrode. The RHE is thus associated with a *reduction* electrode reaction. Thus, for the cell of Fig. 2.3a, reduction occurs at the right hand electrode. We can write the reaction per faraday of electricity passed (ie RHE/F or LHE/F):

$$\text{RHE}/F \qquad M_2^+ + e \longrightarrow M_2$$

By an analogous argument, the electrons for current flow in the external circuit must derive from the electrode reaction at the left hand electrode (LHE), ie *oxidation* must be the spontaneous reaction here. Thus, oxidation occurs at the left hand electrode. We can write the reaction per faraday as:

$$\text{LHE}/F \qquad M_1 \longrightarrow M_1^+ + e$$

The overall cell reaction (OCR) is obtained by adding the two electrode reactions, once these have been written down for the same number of faradays. In this case we are dealing with one faraday in each electrode reaction so that we can add the two electrode reactions directly:

$$\text{OCR/F} \qquad M_1 + M_2^+ \rightleftharpoons M_1^+ + M_2$$

This reaction is the spontaneous reaction of the galvanic cell once the electrical circuit between the two electrodes is completed externally. The cell reaction may not always be a chemical reaction, eg it can be a concentration change of a cell constituent.

∏   Two Class I electrodes are constructed, one a copper electrode and the other a silver electrode. On joining the half cells together we find experimentally that the silver electrode has the higher potential. Deduce the spontaneous electrode and cell reactions.

Your answer should be:

$$\text{LHE}/2F \quad Cu(s) \qquad\qquad \longrightarrow \quad Cu^{2+} + 2e$$
$$\text{RHE}/F \quad Ag^+ + e \qquad\quad \longrightarrow \quad Ag(s)$$
$$\text{OCR}/2F \quad Cu(s) + 2\,Ag^+ \quad \rightleftharpoons \quad Cu^{2+} + 2\,Ag(s)$$

Because the silver electrode is the positive electrode of the cell, reduction must occur spontaneously here. The minimum reaction is a one electron transfer, ie a one faraday process for a mole of silver ions.

Copper is the left hand electrode and must have an oxidation reaction occuring at it. Since the copper(II) ion is doubly charged, the reaction requires two electrons.

The imbalance in the number of electrons in the two reactions must be eliminated by doubling the RHE reaction before the electrode reactions are added.

Note that the OCR is the oxidation of copper by the silver ion.

Convention 2 The emf of the cell is positive.

The significance of this statement is that the relationship between cell emf and the cell reaction free energy change (Eq. 2.1a) yields a negative free energy change if E is positive. The overall cell reaction is therefore a spontaneous process as discussed previously.

Convention 3 All electrodes are considered as *reduction electrodes* with an electrode potential (symbol E) which is a reduction potential. We must consequently think of all electrodes in principle, as right hand electrodes. However, one electrode in the cell must be the LHE. How therefore do we treat this electrode? The clue is that the only difference between the electrode reaction when an electrode is moved from the right hand position to the left hand position is a reversal in the electrode reaction and the sign of the potential. Let us see how this happens in the case of the silver electrode.

As the RHE the reduction electrode reaction is:

$$Ag^+ + e \longrightarrow Ag(s)$$

and the electrode potential is its reduction potential,

$$E(Ag^+, Ag).$$

As the LHE the electrode reaction is oxidation:

$$Ag(s) \longrightarrow Ag^+ + e$$

the electrode potential is $-E(Ag^+, Ag)$.

On the basis of this argument the emf of any cell is given by the equation:

$$E = E(\text{RHE}) - E(\text{LHE}) \qquad (2.3a)$$

Convention 4 A single electrode reduction potential is equal to the emf of the cell formed with this electrode as the RHE and the standard hydrogen electrode (SHE) as the LHE.

The SHE is arbitrarily assigned a potential of zero, ie

$$E^{\ominus}(H^+, H_2) = 0 \text{ V.}$$

This convention applies at all temperatures. Fig. 2.3b lists some of the important standard single electrode potentials.

You should remember that the word 'standard' implies that the reactants and products in the electrode reaction are at unit activity. If we wish to calculate the electrode potential for any other condition of the electrode reactants and products, then it is a relatively simple matter to do so as we will see in a later section.

You will notice that the electrode potential can be positive or negative with respect to the SHE. Electrodes with a more positive potential must have an electrode reaction which is more reducing than that for the hydrogen ion.

$E^{\ominus}/V$	Electrode Reaction		
2.65	$F_2 + 2e$	$\longrightarrow$	$2\,F^-$
1.77	$H_2O_2 + 2\,H^+ + 2e$	$\longrightarrow$	$2\,H_2O$
1.61	$Ce^{4+} + e$	$\longrightarrow$	Ce^{3+}
1.36	$Cl_2 + 2e$	$\longrightarrow$	$2\,Cl^-$
1.33	$Cr_2O_7^{2-} + 14\,H^+ + 6e$	$\longrightarrow$	$2\,Cr^{3+} + 7\,H_2O$
1.23	$O_2 + 4\,H^+ + 4e$	$\longrightarrow$	$2\,H_2O$
0.799	$Ag^+ + e$	$\longrightarrow$	Ag
0.789	$Hg_2^{2+} + 2e$	$\longrightarrow$	$2\,Hg$
0.771	$Fe^{3+} + e$	$\longrightarrow$	Fe^{2+}
0.682	$O_2 + 2\,H^+ + 2e$	$\longrightarrow$	H_2O_2
0.536	$I_2 + 2e$	$\longrightarrow$	$2\,I^-$
0.337	$Cu^{2+} + 2e$	$\longrightarrow$	Cu
0.268	$Hg_2Cl_2 + 2e$	$\longrightarrow$	$2\,Hg + 2\,Cl^-$
0.222	$AgCl + e$	$\longrightarrow$	$Ag + Cl^-$
0.000	$2\,H^+ + 2e$	$\longrightarrow$	H_2
−0.403	$Cd^{2+} + 2e$	$\longrightarrow$	Cd
−0.41	$Cr^{3+} + e$	$\longrightarrow$	Cr^{2+}
−0.763	$Zn^{2+} + 2e$	$\longrightarrow$	Zn
−2.71	$Na^+ + e$	$\longrightarrow$	Na

Fig. 2.3b.   *Standard electrode potentials (298 K)*

∏   Answer true or false to the following statements which are based on the values listed in Figs. 2.2c and 2.3b.

(a) The cell formed from a standard silver electrode and the SHE has the silver electrode as the RHE and a cell emf of 0.799 V.

(b) The cell formed from a standard Fe^{3+}, Fe^{2+} redox electrode and the saturated calomel electrode at 25 °C has a cell emf of 1.012 V with the calomel electrode as the RHE.

(c) The redox reaction between ions at unit activity:

$$Ce^{4+} + Cr^{2+} \rightleftharpoons Ce^{3+} + Cr^{3+}$$

proceeds spontaneously to the right at 25 °C.

Your answers should be,

    (*a*)   True (*b*)   False (*c*) True

In (*a*) we can see that $E^{\ominus}(Ag^+,Ag) = 0.799$ V.

Since this is a more positive potential than that for the SHE ($E^{\ominus}(H^+,H_2) = 0$ V), the silver electrode is the RHE. From Eq. 2.3a the standard emf of cell, $E^{\ominus} = 0.799 - 0 = 0.799$ V.

The statement in (*a*) is therefore true.

For (*b*), we obtain from the two tables:

$$E^{\ominus}(Fe^{3+}, Fe^{2+}) = 0.771 \text{ V and } E(\text{Cal}) = 0.241 \text{ V}$$

The iron couple must be the RHE and the cell emf is:

$$E^{\ominus} = 0.771 - 0.241 = 0.530 \text{ V}.$$

The statement in (*b*) is consequently false.

For (*c*), we need to obtain the two redox standard potentials:

$$E^{\ominus}(Ce^{4+}, Ce^{3+}) = 1.61 \text{ V} \qquad \text{and}$$
$$E^{\ominus}(Cr^{3+}, Cr^{2+}) = -0.41 \text{ V}$$

If these two redox electrodes are joined together in a cell, the cerium redox electrode will form the RHE and the cell reactions will be:

RHE/F	$Ce^{4+} + e$	$\longrightarrow$	Ce^{3+}
LHE/F	Cr^{2+}	$\longrightarrow$	$Cr^{3+} + e$
OCR/F	$Ce^{4+} + Cr^{2+}$	$\longrightarrow$	$Ce^{3+} + Cr^{3+}$

The standard emf is:

$$E^{\ominus} = 1.61 - (-0.41) = 2.02 \text{ V}.$$

Note that the cell reaction is the one quoted in statement (c) and is the spontaneous reaction of the cell. The statement is therefore true.

You can go further and calculate the standard Gibbs (free) energy change for the reaction, (Eq. 2.1a):

$$\Delta G^{\ominus} = -zE^{\ominus}F = -1 \times 2.02 \times 96\,487 = -194.7 \text{ kJ}$$

2.3.1.   Linear Cell Format

In dealing with electrochemical cells it is convenient to be able to write them down in a simple manner and not as a diagram (Fig. 2.3a). A common method of doing this is the linear cell format for which the following set of rules applies.

Rule 1: the electrode/electrolyte interface is indicated by a vertical line, eg:

$$\text{AgNO}_3(0.1 \text{ mol dm}^{-3}) \mid \text{Ag}$$

$$\text{Fe}^{3+}, \text{Fe}^{2+} \mid \text{Pt}$$

Rule 2: the components of an electrode or electrolyte are separated by commas, eg:

$$\text{NaCl}(1.0 \text{ mol dm}^{-3}) \mid \text{AgCl, Ag}$$

$$\text{NaCl}(1.0 \text{ mol dm}^{-3}), \text{NaF}(0.1 \text{ mol dm}^{-3}) \mid \text{AgCl, Ag}$$

Rule 3: a liquid–liquid junction is indicated by a double vertical line, eg:

$$\text{Cu} \mid \text{Cu(NO}_3)_2 \ (a_1) \parallel \text{Ag(NO}_3)_2 \ (a_2) \mid \text{Ag}$$

If the liquid junction is formed using a salt bridge, the electrolyte used in this bridge is indicated between the two pairs of vertical lines, eg:

$$Cu \mid Cu(NO_3)_2 \ (a_1) \parallel KNO_3 \parallel AgNO_3 \ (a_2) \mid Ag$$

Rule 4: the concentrations or activities of the electrode or electrolyte components and their physical state and the solvent should be indicated in parenthesis after the particular component.

This is shown in the examples above. However, it is difficult to cover all conceivable situations with any set of rules and a certain amount of understanding and flexibility is needed in applying the rules. Thus it is unusual to state the solvent water since solutions are commonly aqueous. Cell components which are obviously solid often have the (s) omitted.

∏   Indicate whether the following statements are true or false.

    (*a*) The cell Ag, AgCl $\mid$ HCl(a) $\mid$ H$_2$(p), Pt has the spontaneous reaction:

$$Ag + HCl(a) \longrightarrow AgCl + 0.5\,H_2(p)$$

    (*b*) A cell constructed from a chlorine electrode (gas pressure = 1 atmosphere) dipping into a 0.1 mol dm^{-3} solution of NaCl and a saturated potassium chloride calomel electrode has the linear cell format:

$$Hg, Hg_2Cl_2 \mid KCl(sat.) \parallel NaCl \mid Cl_2(p = 1 \text{ atmos})(Pt)$$

The answers are (*a*) True and (*b*) False.

In (*a*) we must assume that the RHE is the positive electrode and that reduction occurs here:
$$RHE/F \quad H^+(a) + e \longrightarrow 0.5\,H_2(p)$$

At the LHE oxidation occurs:

$$LHE/F \quad Ag + Cl^- \longrightarrow AgCl + e$$

By adding these two equations we have:

$$\text{OCR/F}\quad \text{Ag} + \text{HCl}(a) \longrightarrow \text{AgCl} + 0.5\,\text{H}_2(p)$$

This is the stated reaction and the statement is therefore true.

In (b) we must first establish which electrode is the positive electrode. From Figs. 2.2c and 2.3b, we see that the chlorine electrode has the higher electrode potential and is thus the RHE as given in the statement. The format however differs from that recommended in two respects, (i) the chlorine electrode should read $\text{Cl}_2(p = 1$ atmos), Pt and (ii) the electrolyte concentration is missing.

The statement is false.

SAQ 2.3a

A galvanic cell is constructed from two redox electrodes, one involving titanium(IV) and titanium(III) ions and the other involving cerium(IV) and cerium(III) ions. If the ions are all at unit activity and you are told that $E^{\ominus}(\text{Ti}^{4+}, \text{Ti}^{3+}) = -0.04$ V.

(i)   Decide which electrode will be the positive electrode.

(ii) Write down the cell linear format.

(iii) Write down the electrode and cell reactions.

(iv) Calculate the free energy change for the spontaneous cell reaction.

SAQ 2.3a

SUMMARY AND OBJECTIVES

Summary

A galvanic cell must be written down according to the IUPAC conventions. These place the electrode with the higher potential as the right hand electrode and the spontaneous electrode reaction here is reduction. The cell emf is positive. All electrode potentials are referred to the standard hydrogen electrode which is assumed to have a potential of zero. The written form of a cell shows the electrode/electrolyte interface as a vertical line, a liquid–liquid junction as two vertical lines and any electrode and cell components separated by commas. The physical states of all components must be indicated.

Objectives

You should now be able to:

- state and apply the IUPAC conventions for galvanic cells;

- explain the rules for linear formatting of cells;

- write a galvanic cell from a knowledge of the two standard electrode potentials.

2.4. THE NERNST EQUATION

The Nernst equation relates the cell emf or electrode potential to the activities of substances participating in the cell or electrode reaction respectively. The Nernst equation therefore is the cell or electrode response equation.

We will briefly look at the origin of the Nernst equation so that we can later write it down for any galvanic cell. We can represent a cell or electrode reaction by a perfectly general equation,

$$aA + bB \rightleftharpoons cC + dD$$

example 1 $Cu^{2+} + 2e \longrightarrow Cu(s)$

(here $a = 1$, A $= Cu^{2+}$, $b = 2$, B $= e$, $c = 1$, C $=$ Cu, $d = 0$, $z = 2$)

example 2 $Ag(s) + HCl = AgCl(s) + 0.5\,H_2$

(here $a = 1$, A $=$ Ag, $b = 1$, B $=$ HCl, $c = 1$, C $=$ AgCl, $d = 0.5$, D $= H_2$, $z = 1$)

The change in free energy for this reaction is:

$$\Delta G = cG_C + dG_D - aG_A - bG_B$$

Since Eq. 1.2e will apply to each chemical species in this equation, we can change the above to:

$\therefore \qquad \Delta G = \Delta G^\ominus + RT \ln \dfrac{\{a^c(C)\ a^d(D)\}}{\{a^a(A)\ a^b(B)\}}$

This equation is converted into one involving emf using $\Delta G = -zEF$ (Eq. 2.1a):

$$E = E^\ominus + (RT/zF) \ln \dfrac{\{a^a(A)\ a^b(B)\}}{\{a^c(C)a^d(D)\}}$$

It is more common to write this equation involving logarithms to the base 10, ie,

$$E = E^{\ominus} + (2.303\ \mathrm{R}T/zF)\ \lg \frac{\{a^{a}(\mathrm{A})\ a^{b}(\mathrm{B})\}}{\{a^{c}(\mathrm{C})\ a^{d}(\mathrm{D})\}} \quad (2.4a)$$

This equation is the NERNST equation.

The pre-logarithmic factor $2.303\ \mathrm{R}T/F$ is known as the Nernst factor and its value is listed in Fig. 2.4a, at several temperatures. It is interesting to note that (*i*) the electrode response changes approximately 60 mV for a 10-fold change in the analyte activity, and (*ii*) the response depends on the analyte activity and not its concentration, although these two quantities are not very different in very dilute solution. The ideal response line for an electrode is shown in Fig. 2.4b. Many real electrodes show response lines with a smaller slope than the ideal Nernstian slope.

$T/°C$	Nernst factor/mV
0	54.21
15	57.17
20	58.16
25	59.15
30	60.13
40	62.12

Fig. 2.4a. *The Nernst factor*

We will now apply Eq. 2.4a to the two examples given earlier in this section. Remember that the activity of a pure substance and an electron is unity and therefore they do not appear in the response equation. For example 1 we have:

$$E(Cu^{2+}, Cu) = E^{\ominus}(Cu^{2+}, Cu)$$
$$+ (2.303\ \mathrm{R}T/2F)\ \lg a(Cu^{2+}) \quad (2.4b)$$

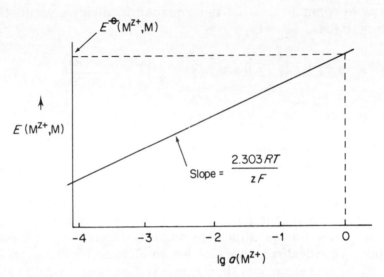

Fig. 2.4b. *Ideal (Nernstian) electrode response for a Class I electrode. $M^{z+} + ze \rightleftharpoons M$*

If you turn back to Eq. 2.2b you will see that this equation above is exactly equivalent to it.

Applying Eq. 2.4a to example 2 we have:

$$E = E^{\ominus} + (2.303\ RT/F)\ \lg \frac{\{a(H^+)\ a(Cl^-)\}}{\{p^{\frac{1}{2}}(H_2)\}}$$

The cell emf is thus dependant upon the hydrogen gas pressure as well as the activities of the hydrogen and chloride ions.

∏   (*a*) Write down the electrode response equation for the redox electrode Ce^{4+}, Ce^{3+} | Pt and state the conditions under which the electrode potential will equal its standard value.

(*b*) Deduce the change in the emf of a cell (at 298 K) whose reaction is:

$0.5\ H_2 + 0.5\ Cl_2 \rightleftharpoons HCl$

if the concentration of HCl in the cell is decreased 10-fold. (Neglect activity coefficients).

The answer for (*a*) is:

$$E(Ce^{4+}, Ce^{3+}) = E^{\ominus}(Ce^{4+}, Ce^{3+})$$
$$+ (2.303\,RT/F)\,\lg\,\{a(Ce^{4+})/a(Ce^{3+})\}$$

The value of E will equal $E^{\ominus}$ when the ratio of the cerium ion activities is unity. In dilute solutions this will be when their concentrations are equal.

The electrode reaction is:

$$Ce^{4+} + e \longrightarrow Ce^{3+}$$

Comparing this to the general reaction above we see that $a = 1$, $A = Ce^{4+}$, $b = 1$, $B = e$, $c = 1$, $C = Ce^{3+}$, $d = 0$, $z = 1$. Putting these values into Eq. 2.4a gives the equation quoted above.

For (*b*) the answer is 118 mV.

The cell Nernst equation can be written down after comparing our cell reaction with the general reaction ($a = 0.5$, $A = H_2$, $b = 0.5$, $B = Cl_2$, $c = 1$, $C = H^+$, $d = 1$, $D = Cl^-$, $z = 1$). You must remember that strong electrolytes in water will be, or at least are usually assumed to be, fully dissociated. Thus we must consider HCl as its two dissociated ions, H^+ and Cl^-.

The response equation is,

$$E = E^{\ominus} + (2.303\,RT/F)\,\lg\,\frac{\{p^{\frac{1}{2}}(H_2)\,p^{\frac{1}{2}}(Cl_2)\}}{a(H^+)\,a(Cl^-)}$$

Upon changing the HCl concentration, only the terms $a(H^+)$ and $a(Cl^-)$ will change. If we call the initial HCl concentration c then the diluted solution will be $0.1\,c$ in concentration and the difference in cell emf's will be,

$$E(0.1\,c) - E(c) = (2.303\,RT/F)\,\lg\,\{c^2/(0.1\,c)^2\}$$
$$= 59.1 \times \lg\,(100)$$
$$\therefore \quad E = 118.3 \text{ mV}$$

SAQ 2.4a   A galvanic cell is devised to monitor the cadmium ion activity in cadmium sulphate solutions
at 25 °C. The cell consists of a cadmium indicator electrode and a saturated calomel reference
electrode. Deduce for this cell:

(i)   the cell linear format;
(ii)   the electrode and cell reactions;
(iii) the cell response equation;
(iv)   the cell emf when the analyte is 0.1 mol
         dm^{-3} cadmium sulphate solution ($y\pm$ =
         0.150).

(use Figs. 2.2c, 2.3b, 2.4a as necessary)

SUMMARY AND OBJECTIVES

Summary

An electrode or cell reaction can be formulated as:

$$aA + bB \rightleftharpoons cC + dD$$

and the corresponding Nernst equation is:

$$E = E^{\ominus} + (2.303\,RT/F)\ \lg \frac{\{a^a(A)\ a^b(B)\}}{\{a^c(C)\ a^d(D)\}}$$

The Nernst factor $2.303\,RT/F$ has values which are tabulated.

Objectives

You should now be able to:

• formulate the Nernst equation for an electrode reaction and for a cell reaction;

• calculate the emf of a cell from standard electrode potentials and concentrations of reacting species.

2.5. THE LIQUID JUNCTION

In our discussions so far we have ignored any electrical potential which might arise in a galvanic cell from the junction between two liquids. At this boundary, as indeed at any boundary between two phases, a potential difference does occur and is known variously as the *liquid junction potential*, the *boundary potential* or the *diffusion potential*. We shall refer to it as the liquid junction potential and its abbreviation, ljp. The magnitude of the ljp depends upon the concentrations and the nature of the ions on the two sides of the junction.

In order to understand the origin of the ljp we will consider a concentration cell, ie one which develops its emf from the difference in concentration between the two half-cell electrolytes.

$$\text{Ag} \mid \text{AgNO}_3 \ (c_1) \parallel \text{AgNO}_3 \ (c_2) \mid \text{Ag} \quad c_2 > c_1$$

The two silver nitrate solutions could be separated by means of some porous element, eg a glass frit or a porous ceramic. As soon as the junction between the two silver nitrate solutions is formed, silver and nitrate ions diffuse across it. More ions will pass from right to left than left to right because of the higher concentration on the right hand side. Ions diffuse at velocities which are proportional to their mobilities (Fig. 1.3a). Therefore the nitrate ion will initially diffuse faster than the silver ion leading to a charge distribution across the boundary as below,

Zero time   Short time later

You can see that a separation of some of the positive and negative charges has occurred and a potential difference has formed across the junction. This potential difference will not continue to increase because it will itself increase the speed of the silver ions and decrease the speed of the nitrate ions due to Coulombic forces. Consequently, and very rapidly, a steady-state potential difference is established and this is known as the liquid junction potential. In this case the ljp is positive: it is symbolised by E_L.

By thermodynamic arguments it can be shown that for this situation, where the liquid junction is formed between the same electrolyte at different concentrations:

$$E_L = (t_- - t_+)(2.303 \, RT/F) \lg \{(a_2)/(a_1)\} \qquad (2.5a)$$

Transport numbers do change with electrolyte concentration and temperature to some extent, but they alter more significantly if the solvent is varied. Thus the potassium ion in aqueous KCl has a transport number of approximately 0.49 but in dimethyl formamide it falls to 0.36. A liquid junction using KCl in this solvent would develop a significant ljp.

So far our consideration has been of a somewhat unusual junction, ie that between solutions of the same electrolyte at different concentrations. A more general situation is where two different electrolytes at different concentrations form the liquid junction. This is not an easy case to treat theoretically and it is consequently difficult to accurately predict the value of the ljp.

How then do we cope with such a situation in the practice of electroanalytical chemistry? We approach the problem in one of two ways. Firstly, a *salt bridge* can be employed between the two electrolyte solutions to minimise the ljp and hopefully eliminate it in some cases. Secondly, we attempt to keep the ljp constant during the calibration and analysis stages of a determination of an analyte.

A salt bridge is interposed between the two half-cell solutions as shown in Fig. 2.5a. It consists of a concentrated solution of an electrolyte whose ions have transport numbers near to 0.5, eg KCl.

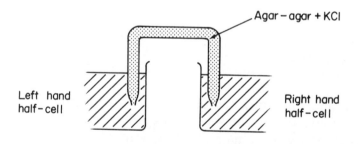

Agar – agar + KCl

Left hand half-cell

Right hand half-cell

Fig. 2.5a. *Salt bridge*

A concentrated solution is chosen so that the diffusion potential across the junction is dominated by the bridge electrolyte. Fig. 2.5b shows the effect of electrolyte concentration and type on the value of the ljp.

| Solution 1 | | Solution 2 | | E_L/mV |
Electrolyte	Conc. /mol dm^{-3}	Electrolyte	Conc. /mol dm^{-3}	
HCl	1.0	KCl	3.5	16.6
HCl	0.1	KCl	3.5	3.1
HCl	0.01	KCl	3.5	1.4
KCl	1.0	KCl	3.5	0.2
KCl	0.1	KCl	3.5	0.6
KCl	0.01	KCl	3.5	1.0
HCl	1.0	KCl	0.1	56.2
HCl	0.01	KCl	0.1	9.3
LiCl	0.1	KCl	0.1	−8.9
NaCl	0.1	KCl	0.1	−6.4

Fig. 2.5b.   *Liquid junction potentials (H_2O, 298 K)*

An alternative to this type of salt bridge is the one made from agar–agar gel containing some electrolyte. Whichever type of bridge is used, it must be remembered that there is a constant outflow of bridge electrolyte into the half-cell solutions. This has been discussed earlier in conjunction with the calomel electrode.

Electrolytes which are commonly used in salt bridges are, KCl, K_2SO_4, KNO_3 and NH_4NO_3 in aqueous systems and NH_4ClO_4 and lithium trichlorethanoate in non-aqueous situations.

In many analytical procedures the ljp is maintained throughout as constant as possible by adding an electrolyte to the analyte solution. This is chosen to dominate the ionic transport across the junction from the analyte side. The electrolyte would have ionic transport numbers within 5% of one another and the electrolyte concentration must be much higher than the analyte concentration which is the changing factor during calibration and analysis. An electrolyte chosen from the list above is suitable. This same addition to the analyte solution can also serve to fix the ionic strength so that the ionic activity coefficient of the analyte remains constant.

SAQ 2.5a   Calculate the ljp at 25 °C for the junction between each of the following pairs of electrolytes.
Assume that activity coefficients are unity and
that the transport number does not change with
electrolyte concentration.

(*i*) $AgNO_3$ $c_1 = 0.005$ mol dm^{-3}: $c_2 = 0.01$ mol dm^{-3}

(*ii*) KCl $c_1 = 0.005$ mol dm^{-3}: $c_2 = 0.01$ mol dm^{-3}

(*iii*) HCl $c_1 = 0.001$ mol dm^{-3}: $c_2 = 0.01$ mol dm^{-3}

SUMMARY AND OBJECTIVES

Summary

A liquid junction potential originates from the different rates of diffusion of the ions across a liquid junction in the galvanic cell. The
prediction of the magnitude and sign of the ljp is not generally possible. In any analytical situation, the ljp is either minimised by using
a salt bridge or maintained constant by having a high electrolyte
concentration on the analyte side of the liquid junction.

Objectives

You should now be able to:

● explain the nature and practical treatment of the liquid junction in a galvanic cell.

2.6. EMF MEASUREMENT

The emf of a galvanic cell should be measured when zero current is flowing through it. In this state the cell is at equilibrium and no chemical reaction is occuring within it. It is however, impossible to measure the emf without a small current flowing. We must therefore ensure that this small current does not significantly alter the value of the emf measured

Three types of measuring device are commonly used for emf measurement – the potentiometer has been widely used for galvanic cells of low internal resistance ($<10^3 \Omega$) but is being displaced by the digital voltmeter (dvm) which can be used for cells of somewhat higher internal resistance. For high resistance cells ($>10^9 \Omega$) such as those involving the glass electrode, a pH meter or ion-selective meter must be used.

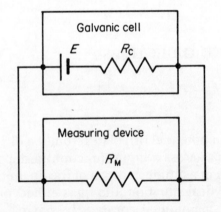

Fig. 2.6a.   *Emf measuring circuit*

The potentiometer will be dealt with separately later; for the moment we will consider the dvm and the pH meter which are connected to the cell as shown in Fig. 2.6a. Included in the diagram are the internal resistance of the cell (R_c) and its emf (E) and the input impedance of the meter measuring device (R_m). The diagram shows that the cell emf is divided between the cell itself (E_c) and the measuring device (E_m). The latter only 'sees' and measures the voltage across its terminals and therefore records the emf as E_m. By analysis of the circuit we can show that:

$$E/E_m = R_c/R_m + 1 \qquad (2.6a)$$

An error of 0.1% in the measured emf, ie $E/E_m = 1.001$, results from a cell resistance of one thousandth the value of the meter input impedance, ie $R_c/R_m = 0.001$.

This discussion highlights the need for us to be aware of the electrode and solution resistances as well as the meter input impedance before making any measurements of cell emf. Solution resistances are increased dramatically on adding non-aqueous solvents to the galvanic cell.

A dvm typically has an input impedance of 10 MΩ allowing the emf of galvanic cells up to a resistance of $10^4 \Omega$ to be measured accurately. We could not use glass electrodes with such a device since their resistance is usually greater than 1 MΩ. The pH meter having an input impedance of 10^6 MΩ can be used for cells with resistances up to 10^3 MΩ.

It is instructive to pause a moment to reflect on the magnitude of the current flowing in the circuit when using a pH meter. Since the total resistance of the circuit is approximately 10^6 MΩ, if the cell emf is 1 V the current flowing will be 10^{-12} A. A change of emf of 1 mV will thus correspond to a change in current of 10^{-15} A. It is little wonder that such delicate measurements are readily upset by stray currents in the circuit and by external influences. Great care must be exercised in shielding electrode leads and in having a good common earth for the measurement system.

ⅠⅠ   A galvanic cell including a glass electrode has a resistance of 1 MΩ. A pH meter having an input impedance of 10^6 MΩ, shows an emf for the cell of 500 mV. What reading would a digital voltmeter record for the cell emf if its input impedance is $10^7 \Omega$?

Your answer should be 454.5 mV.

The first task which you must do is to check that the pH meter is recording the true emf of the cell. From Eq. 2.6a:

$$E/E_m = R_c/R_m + 1 = 10^6/10^{12} + 1$$

This ratio is thus shown to be extremely close to unity. The pH meter therefore records the true cell emf.

Turning our attention to the dvm:

$$E/E_m = 500/E_m = R_c/R_m + 1 = 10^6/10^7 + 1 = 1.1$$

$$\therefore \quad E_m = 500/1.1 = 454.5 \text{ mV}.$$

A 10% lower voltage is shown by the dvm demonstrating the serious nature of the errors which can arise in choosing an unsuitable measuring device for cell emf.

In the use of electrochemical cells for analytical purposes, the accuracy of concentration measurement depends upon the accuracy to which we can measure the cell emf. If we consider the cell emf to be related to the analyte concentration as:

$$E = E^{\ominus} + (RT/zF) \ln c(A)$$

then it can be shown that the percentage error in concentration (P_c) is given by:

$$P_c = 100\Delta E/\{RT/zF\} \tag{2.6b}$$

At 25 °C, P_c is approximately $4z \, \Delta E$ where ΔE is in mV. For a reaction involving one electron transfer at the electrodes then an error in cell emf measurement of 1 mV, leads to a concentration error of 4%.

2.6.1. The Potentiometer

The potentiometer consists of a long resistance wire of very even quality and dimensions. A constant voltage source, eg a battery, drives a steady small current through the wire which consequently has a potential across any length of it which is proportional to that length. Referring to Fig. 2.6b we have:

potential difference across AC $\propto$ length AC
potential difference across AD $\propto$ length AD

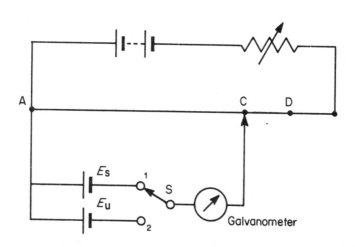

Fig. 2.6b. *The potentiometer*

The wire is often manufactured from platinum or platinum/irridium and has a resistance of $10^3 \Omega$.

The potentiometer is a precise variable voltage source and is used to measure cell emf by the Poggendorf Compensation Method. In this method, the wire is first calibrated by connecting a reference voltage source (E_s), eg a Weston cadmium cell, through a galvanometer (a current measuring device) between the points A and a point C on the wire, which is the point where the galvanometer shows no current to be flowing through it. In this null balance condition we know that the cell emf (E_s) exactly balances the wire emf represented by its length AC:

$$E_s \propto \text{length AC} \qquad (2.6c)$$

If we now connect the cell of unknown emf through the galvanometer, by switching S to position 2, a new balance point at D leads to:

$$E_u \propto \text{length AD} \qquad (2.6d)$$

The ratio of Eqs. 2.6d and 2.6c gives,

$$E_u / E_s = \text{AD/AC}$$

and from this relationship the emf of the unknown cell is obtained.

The potentiometer, with proper attention to detail in its construction and use, can be used to make very precise voltage measurements. The normal laboratory form of this device should not be used for cells whose resistance exceeds $10^4 \Omega$.

2.6.2. The Standard Cell

In the previous section, the need for a primary standard of voltage was established. Commercial electronic sources are available but a common laboratory standard is the Weston cadmium cell. The desired characteristics of such a source are that it should be a very stable chemical system containing readily purified chemical substances. The cell should also have a low temperature coefficient of emf. The Weston cell is depicted in Fig. 2.6c. The overall cell reaction is:

$$Cd + Hg_2SO_4 + 8/3\,H_2O \rightleftharpoons CdSO_4.8/3\,H_2O + 2\,Hg$$

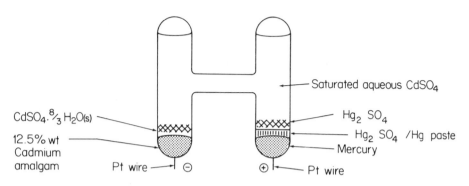

Fig. 2.6c. *The Weston cadmium cell*

The cell emf at 20 °C is 1.018300 V if the cadmium sulphate is saturated at 20 °C. The temperature coefficient is very low at $-40 \, \mu$ V K^{-1} and this allows the cell to be used for many purposes without the need to be thermostatted. In use the cell must be treated with care, a current of no more than approximately 50 μA should be drawn, and it should not suffer vibration or be subjected to temperatures outside the range 4 to 40 °C.

SAQ 2.6a You are provided with a potentiometer having a galvanometer of sensitivity of 1 μA per mm deflection and a resistance of 100Ω. How accurately are you likely to be able to measure the emf of a cell whose internal resistance is 1900Ω? Is this an acceptable level of accuracy for 1% accuracy in concentration?

SUMMARY AND OBJECTIVES

Summary

The emf of a galvanic cell can be measured by means of, a potentiometer (if the cell resistance is less than $10^3 \Omega$), a digital voltmeter
(if the cell resistance is less than $10^4 \Omega$), or a pH meter (which must
be used if the cell resistance is greater than $10^9 \Omega$). The calibration
of any voltage measuring device can be achieved using a standard
cell which has a known, reproducible and stable emf.

Objectives

You should now be able to:

- explain the practice of emf measurement;

- describe the structure and properties of a standard cell.

2.7. pH AND ITS MEASUREMENT

Several electrodes can be used as indicators for the hydrogen ion
activity in solution, eg the hydrogen electrode, the glass electrode.
The latter electrode is the most widely used laboratory electrode.
A galvanic cell for the measurement of hydrogen ion activity can
therefore be constructed by combining the glass electrode with a
reference electrode, usually the saturated calomel electrode.

$$\text{glass} \mid H^+ \text{ solution} \parallel \text{sat.KCl} \mid Hg_2Cl_2, \; Hg$$
$$\text{(analyte)}$$

The emf of this cell will be given by,

$$E \; = \; E(\text{cal}) \; - \; E'(\text{glass}) \; + \; E_L \; + \; (2.303 \, RT/F) \, \text{pH} \quad (2.7a)$$

Note that we have now included the ljp in the cell response equation to take account of the liquid junction potential at the analyte/saturated KCl boundary. If we use the formal definition of pH, ie pH = $-\lg a(H^+)$, then we cannot deduce the pH-value of the analyte from the cell emf using Eq. 2.7a unless the ljp is known. This is not usually the case.

For this and other reasons, the formal definition of pH has been replaced by an operational definition based on a cell of the general form just discussed. To illustrate the definition, we will consider the measurement of the cell emf when firstly, it contains a buffer of known pH, and secondly, when it contains the analyte solution of unknown pH. If the pH values are pH_s and pH_u and the emf values, E_s and E_u respectively, we can obtain from Eq. 2.7a:

$$E_s = E(\text{cal}) - E'(\text{glass}) + E_L + (2.303\,RT/F)\,pH_s$$

$$E_u = E(\text{cal}) - E'(\text{glass}) + E_L + (2.303\,RT/F)\,pH_u$$

We must now assume that the ljp does not change on changing the buffer solution for the analyte solution and this allows us to subtract the two equations. On rearrangement, we obtain:

$$pH_u = pH_s + (E_u - E_s)/(2.303\,RT/F) \qquad (2.7b)$$

You may have noticed a weakness in the logic of the above argument: it is that we knew the pH of a buffer solution before we had even defined pH. The American National Bureau of Standards (NBS) has proposed a set of standard buffer solutions whose pH value has been established from the emf of cells without a liquid junction. Three of these buffer solutions are given in Fig. 2.7a. The operational definition of pH accepts the pH of these buffer solutions as correct. More complete lists of the NBS buffers are found elsewhere (Bates, 1973).

For the pH range 2 to 12 and for low ionic strength solutions (<0.1 mol dm^{-3}), the measured pH using the method indicated should be within 0.02 of the correct value. Outside these ranges, errors in excess of 0.1 are possible.

	Buffer pH		
Temp/°C	1	2	3
0	4.003	6.984	9.464
10	3.998	6.923	9.332
20	4.002	6.881	9.225
25	4.008	6.865	9.180
30	4.015	6.853	9.139
40	4.035	6.838	9.068
50	4.060	6.833	9.011

Buffer compositions:

1. 0.05 mol dm^{-3} potassium hydrogen phthalate
2. 0.025 mol dm^{-3} in potassium dihydrogen phosphate and disodium hydrogen phosphate.

3. 0.01 mol dm^{-3} borax.

Fig. 2.7a.   *NBS buffer solutions*

∏   Some pH measurements are made at 35 °C. The pH meter is first calibrated with a 0.05 mol dm^{-3} potassium hydrogen phthalate buffer solution. An enzyme-containing solution then shows a pH of 4.65. What is the change in cell emf on replacing the enzyme solution with the buffer solution.

The answer is 38.21 mV. You should be considering Eq. 2.7b here. The Nernst factor at 35 °C can be interpolated from Fig. 2.4a as 61.13 mV. The pH of the buffer can be deduced from Fig. 2.7a as 4.025. Putting these two values into Eq. 2.7b we have:

$$4.65 = 4.025 + (E_u - E_s)/61.13$$

$$\therefore \quad (E_u - E_s) = 0.625 \times 61.13 = 38.21 \text{ mV}$$

We will conclude this discussion on pH measurement, with a brief description of the use of controls found on a pH meter.

To set up a pH meter for use, the electrodes must first be placed in a buffer solution of known pH which is similar to that expected for the analyte solution. The pH meter has a *buffer adjust* control which allows the pH reading to be adjusted to the correct value. The cell temperature is important for not only does the buffer pH change with temperature but the cell response does also. The latter effect can be accounted for in one of two ways. Firstly, some pH meters have a *temperature* knob which can be set to the cell temperature, the change in cell response being corrected electronically. Secondly, a *temperature compensator* probe can be inserted into the cell electrolyte. This is a temperature-sensitive resistor which is connected to compensating circuits in the pH meter.

The pH meter must be continually checked because of changes which can occur in electrode response particularly with the glass electrode. In addition, if pH measurements are required over a range of values, the pH meter scale must be checked at two points with appropriate buffer solutions. If the pH range is incorrect some meters have a *slope* knob which can be altered to correct this. A change in the electrode response is experienced with age and misuse of the electrode.

SAQ 2.7a

A pH meter shows the following pH readings at 25 °C for two standard buffer solutions of pH 4.00 and 8.90. The correct values are 4.00 and 9.18 respectively for these buffers. What mV output per pH unit is the meter receiving and what should it be receiving?

SAQ 2.7a

SUMMARY AND OBJECTIVES

Summary

The operational definition of pH is based on the emf of a galvanic cell consisting of an indicator electrode for the hydrogen ion and a reference electrode. The emf of the cell (E_u) when it contains the analyte solution as opposed to its emf (E_s) when it contains a buffer solution (pH_s) leads to the pH of the analyte solution (pH_u):

$$pH_u = pH_s + (E_u - E_s)/(2.303\,RT/F).$$

Certain buffer solutions, the NBS buffers, have conventionally accepted pH values. Care must be exercised in pH measurements to allow for the solution temperature and in the choice of a buffer with a pH similar to that of the analyte solution. The pH meter is furnished with controls for temperature, buffer adjustment and sometimes, slope.

Objectives

You should now be able to:

● explain the difference between the formal and operational definitions of pH;

● describe the method of pH measurement using a pH meter.

2.8. POTENTIOMETRY – AN INTRODUCTION

Potentiometry is the measurement of the emf of a galvanic cell which is operating near zero current. Because, as we have seen, this emf is a function of the ionic activities within the cell, it can be used analytically to measure ionic concentrations in titrations, process streams, biological fluids and a multitude of other situations. Potentiometry is, in fact, the most widely used analytical technique. An indicator electrode often yields a reasonable response over an analyte concentration range of 10^{-7} to 1 mol dm^{-3} and it can be employed in both aqueous and non-aqueous environments. The apparatus required to carry out the analyses is both simple and inexpensive and electrodes often remain effective for long periods of time. The electrode is in many ways an ideal sensor since it can be coupled directly into an analogue-to-digital convertor for data acquisition by a computer. Analytical methods involving galvanic cells can therefore be readily automated and computerised.

We will briefly discuss some of the key potentiometric methods and then study the potentiometric titration in detail.

(*a*) *single-point* potentiometry is the direct measurement of an analyte concentration using a cell with an appropriate indicator electrode. In this way the concentration of fluoride in a solution can be measured with a lanthanum fluoride electrode and a mercury(I) sulphate reference electrode. The response of the indicator electrode is first established by direct calibration with solutions of known fluoride ion concentration. The solution pH and ionic strength is kept constant with the TISAB buffer (Total Ionic Strength Adjustment Buffer). This is a mixture of chemicals which fix the solution pH within the operating range of the fluoride electrode and also dominate the ionic strength of the system. The analyte sample is made up in this buffer solution in a similar way to the calibration standards and its emf measured. The analyte concentration can be deduced by comparison with the calibration emf values.

Π   A cell is formed from a lanthanum fluoride electrode and a
       reference electrode. The cell emf for various concentrations
       of fluoride ion, made up in a fixed amount of TISAB solution,
       is tabulated:

$c(F^-)$/mol dm^{-3}	0.001	0.00074	0.00044
E/mV	89.5	82.0	68.5

A mass of 1 g of toothpaste containing sodium monofluo-
rophosphate (smp) was digested in hydrochloric acid and
then made up to 100 cm^3 with water and TISAB buffer of
the same concentration as in the standard solutions. This solution showed a cell response of 74.5 mV. Calculate the %
wt : wt of smp in the toothpaste (smp = Na_2FPO_3, M_r =
144).

The answer is 0.79 % wt : wt.

The cell must have a response equation of the general form of,

$$E = E(\text{ref}) - E^{\ominus}(F^-, F_2) + (2.303\,RT/F)\,\lg c(F^-)$$

The cell emf is consequently directly proportional to $\lg c(F^-)$. From
the cell emf data we can deduce this quantity.

$c(F^-)$/mol dm^{-3}	0.001	0.00074	0.00044
E/mV	89.5	82.0	68.5
$\lg c(F^-)$	-3.00	-3.13	-3.36

A calibration graph is constructed, Fig. 2.8a. and the value of the
fluoride concentration for the toothpaste solution, read from the
graph.

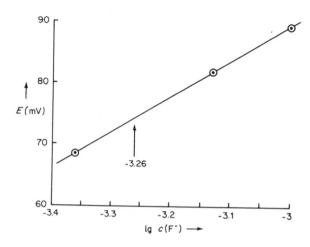

Fig. 2.8a. *Calibration plot of emf against lg c(F⁻)*

For,

$$E = 74.5 \text{ mV}; \lg c(F^-) = -3.26$$

∴ $\quad c(F^-) = 5.49 \times 10^{-4} \text{ mol dm}^{-3}$

Since the solution only totalled 100 cm³, there were 5.49×10^{-5} moles of fluoride in this volume. The compound smp contains one fluorine atom per mole and therefore this quantity is also the moles of smp in 1 g of the toothpaste.

∴ $\quad$ mass of smp in 1 g of toothpaste

$$= 5.49 \times 10^{-5} \times 144 \text{ g}$$
$$= 0.00791 \text{ g}$$

% wt : wt smp in the toothpaste = 0.00791 × 100/1 = 0.79

This technique is used widely for pH measurements and for many other ions. It is simple, rapid and reasonably accurate. If there are interferences from the analyte matrix, a standard additions technique can be adopted instead of the direct calibration method described above.

On-line monitoring methods employ this technique. The response of the indicator electrode must be checked regularly with standard solutions. In addition, since in process streams or river pollution monitoring, there is a flowing liquid, the electrode response in this situation must be ascertained. The response time of the electrode and its change with electrode age may also need to be investigated.

(*b*) *null-point* potentiometry uses a galvanic cell consisting of two identical indicator electrodes, one dipping into the analyte solution of unknown concentration (c_u) and the other dipping into a solution of known concentration (c_k). The two half-cells are connected through a salt bridge. The cell emf is:

$$E = (2.303\,RT/F)\,\lg\{a_u/a_k\} + E_L$$

By adding an excess of an inert electrolyte to the two half-cell solutions, the activity coefficients of the ions being measured should be equal, allowing the ratio of activities in the above equation to be replaced by the ratio of concentrations (c_u/c_k). In addition the ljp at each side of the salt bridge should cancel out, so that $E_L = 0$ in the above equation.

The determination of the analyte concentration now proceeds by adding known amounts of the ion to vary c_k until the cell emf becomes zero. In this condition, $c_k = c_u$.

2.8.1. Potentiometric Titrations

In a potentiometric titration the concentration of an ion which changes during the titration is monitored by an indicator electrode. For example, in an acid-base titration the hydrogen ion concentration is monitored with a glass electrode. As the volume of titrant increases during the titration, the cell emf changes along an 'S' shaped curve as depicted in Fig. 2.8b. The titration end-point is the inflexion point on this curve. The volume of titrant added at the end-point, is equivalent to the amount of the analyte present in the original solution. The titration curve can be analysed further to obtain more fundamental parameters of the system, eg the acid dissociation constant.

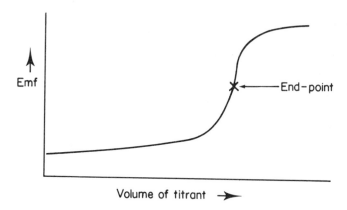

Fig. 2.8b. *Potentiometric titration curve*

A potentiometric titration is not a rapid determination but once the end-point emf has been established, further analyses can be performed rapidly by adding the titrant to the analyte solution until the emf reaches that of the end-point value. These titrations are applicable to coloured, turbid and fluorescent solutions which are difficult to analyse by other methods. In addition, potentiometric titrations are readily performed on non-aqueous solutions and have been used widely for the determination of weak acids and bases.

Redox and precipitation titrations can also be studied potentiometrically. To illustrate the usefulness of the theoretical work on galvanic cells which we have covered, we will look in detail at one such titration – the argentometric halide titration. The approach which we will use could equally well be applied to any other potentiometric titration.

In this titration, silver nitrate (titrant) is added to the chloride (analyte) solution. Silver chloride is precipitated; the amount being governed by the solubility product (K_s) of the silver chloride,

$$AgCl(s) \rightleftharpoons Ag^+ + Cl^-$$

for which:

$$K_s = c(Ag^+)c(Cl^-) \qquad (2.8a)$$

We shall assume a value for K_s of 10^{-10} mol^2 dm^{-6} in water at 25 °C (the actual value is 1.82×10^{-10} mol^2 dm^{-6}). Because the silver ion concentration will increase during the titration, we can monitor this concentration with a silver electrode and use a mercury(I) sulphate electrode as the reference electrode. A typical arrangement of the apparatus is shown in Fig. 2.8c. We can write the cell linear format and electrode and cell reactions and response as:

$$Ag \,|\, analyte\ Cl^-,\ Ag^+ \,\|\, sat.\ K_2SO_4 \,|\, Hg_2SO_4,\ Hg$$

LHE $\quad Ag(s) \qquad\qquad\qquad \longrightarrow \quad Ag^+ + e$

RHE $\quad Hg_2SO_4(s) + 2e \qquad \longrightarrow \quad 2\,Hg(l) + SO_4^{2-}$

OCR $\quad Hg_2SO_4(s) + 2\,Ag(s) \quad \rightleftharpoons \quad 2\,Hg(l) + 2\,Ag^+ + SO_4^{2-}$

$$E \;=\; E(Hg_2SO_4, Hg) \;-\; E^{\ominus}(Ag^+, Ag)$$
$$+\; E_L \;-\; (2.303\,RT/F)\ \lg a(Ag^+)$$

$\therefore \qquad E \;=\; 0.614 \;-\; 0.779 \;+\; E_L \;-\; 0.0592\ \lg a(Ag^+)$

$\therefore \qquad E \;=\; -0.185 \;+\; E_L \;-\; 0.0592\ \lg a(Ag^+) \qquad (2.8b)$

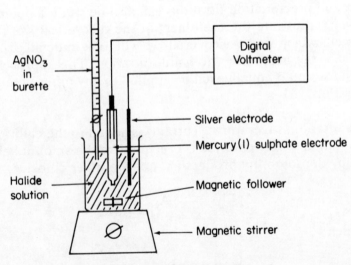

Fig. 2.8c. *Halide titration cell*

We will study the titration for a chloride concentration of 0.01 mol dm^{-3}. The discussion will be simplified by making three assumptions, (*i*) there is no volume change of the analyte solution as the titrant is added (this can be achieved in practice by using a very concentrated titrant solution), (*ii*) the activity coefficient of the silver ion is unity, and (*iii*) the ljp is zero. Applying these three assumptions to Eq. 2.8b we obtain,

$$E = -0.185 - 0.0592 \lg c(Ag^+) \qquad (2.8c)$$

We can now calculate the cell emf at various points in the titration. At the start of the titration, there is no silver ion in the cell and the silver electrode thus has an indeterminate potential. However, as soon as the slightest amount of titrant enters the cell, the silver electrode will respond. Since the solubility product of silver chloride is very small, this salt will precipitate from the beginning of the titration or at least from when the silver concentration exceeds the value given by:

$$K_s/c(Cl^-), \text{ ie } 10^{-8} \text{ mol dm}^{-3}.$$

The effective starting emf of the cell is given then by:

$$E_s = -0.185 - 0.0592 \lg (10^{-8}) = 0.288 \text{ V}$$

As more silver ion is added, more chloride ion is removed and the cell emf falls. At the end-point the silver and chloride ion concentrations in the solution become equal, and both must equal the square root of the solubility product (Eq. 2.8a), ie 10^{-5} mol dm^{-3}. The end point cell emf is thus:

$$E_{ep} = -0.185 - 0.0592 \lg (10^{-5}) = 0.111 \text{ V}$$

After the end-point the excess silver ion will continue to cause the cell emf to fall. Fig. 2.8d shows the complete picture of the change in cell emf during the titration.

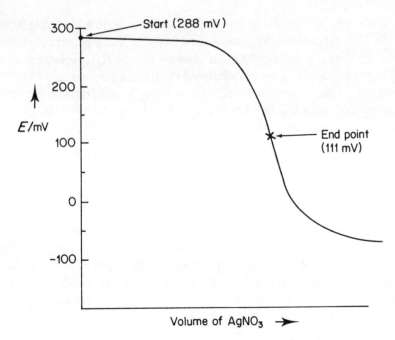

Fig. 2.8d. *Argentometric halide titration*

We have developed quite a detailed view of the titration and this enables us to modify our system to our advantage under any given set of circumstances. For example, we have seen that the change in emf during the titration ($E_s - E_{ep}$), is given by the term:

$$0.0592 \lg \left\{ K_s^{\frac{1}{2}} / c(Cl^-) \right\}.$$

The sensitivity of the titration becomes small when the chloride ion concentration falls below the square root of the solubility product, ie 10^{-5} mol dm^{-3}. If however, we increase the insolubility of the silver chloride by adding ethanol to the cell solution, we can decrease K_s and increase the sensitivity of the titration and make possible the accurate determination of lower chloride concentrations.

The end-point of the titration is the inflexion point on the curve. There are various analytical procedures for determining this particular point. Fig. 2.8e shows a plot of the change of emf (ΔE) for a given small change in titrant volume (ΔV) during the titration.

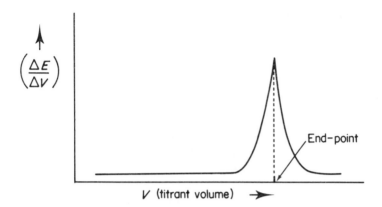

Fig. 2.8e. *Titration curve differential plot*

It is the differential of the titration curve and shows a maximum at the end-point volume. An alternative and simpler procedure is to construct tangents to the curve near the end-point. The bisecting parallel line ab in Fig. 2.8f cuts the titration curve at the end-point.

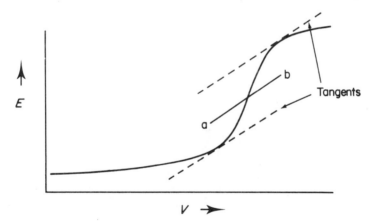

Fig. 2.8f. *End point determination*

The automation of potentiometric titrations has already been alluded to. It can be achieved by having a motor-driven syringe containing the titrant and a Y–t recorder to display the titration curve (Fig. 2.8g). The syringe must be driven synchronously so that the recorder time axis is also the titrant volume axis.

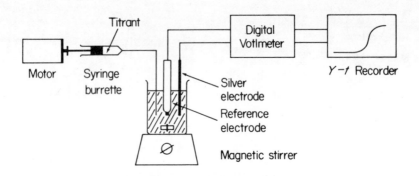

Fig. 2.8g.   *Automated titrator*

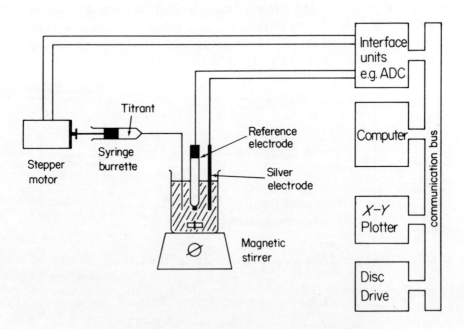

Fig. 2.8h.   *Computerised titration system*

Finally, Fig. 2.8h depicts a computerised potentiometric titration system. The computer software would, upon initiation, switch on the stepping motor adding titrant increments to the galvanic cell. The cell emf is digitised by the analogue-to-digital converter (ADC), after suitable amplification and/or offsetting of the signal, these values are stored in the computer and manipulated to give a printed report on the titration which would include the analyte concentration. The titration curve could be plotted out on the X–Y plotter if necessary either in its normal form or as a first or second derivative curve. Such a system removes much of the manual part of potentiometric titrations and provides an unbiased result of the experiment.

SAQ 2.8a

You have been asked to design an automatic titration system for the determination of the concentration of weak acids in aqueous solution. You have decided to use a cell consisting of glass and saturated calomel electrodes and an automatic titration procedure with sodium hydroxide as the titrant. The cell is known to show zero volts at pH = 7.

Predict the cell emf at 25 °C during the titration of a weak acid ($K = 10^{-5}$) and approximate concentration 0.1 mol dm^{-3}, at:

(*i*)   start of titration,
(*ii*) half neutralisation.

Assume that the activity coefficients of all species are unity, that there is zero ljp and that there is no volume change during the titration.

(Clue: you must consider the acid dissociation equilibrium in a way similar to the use of the solubility product in the text above).

SAQ 2.8a

SUMMARY AND OBJECTIVES

Summary

Single-point potentiometry uses the potential developed at the indicator electrode (versus a reference electrode) to monitor the concentration of an analyte in solution. In contrast, null-point potentiometry employs two indicator electrodes, one in a solution of known and varied concentration of the analyte ion and the other in the solution of unknown concentration of the ion.

The indicator electrode in potentiometric titrations is used to monitor the concentration of the analyte or titrant ion as the titrant volume is increased. This yields a sigmoidal change in emf with titrant volume which can be analysed for the analyte concentration and other physical constants of the chemical system. The cell emf is predictable at all stages of the titration and the dependence of the titration curve shape on the analyte, titrant concentrations and the physical constants of the system can be clearly understood. The apparatus for the manual, automated and computerised titration systems are discussed.

Objectives

You should now be able to:

- describe the theory and practice of single- and null-point potentiometric methods;

- analyse the data from a potentiometric titration;

- describe suitable equipment for undertaking potentiometric titrations.

3. Electrolysis

Overview

In this part the differences between electrochemical cells and electrolysis cells are emphasised and it is established that whereas electrochemical cells form the basis for potentiometry it is electrolysis cells that form the basis for voltammetry.

You are introduced to the principles of three electrode circuitry and potentiostatic control which are essential components of all modern electrolysis based electro-analytical methods, and the importance of good cell design is stressed.

Electrodes, solvents and supporting electrolytes are discussed and here the main objective is to establish that these have to be considered together as a system chosen for a particular application.

Finally the factors affecting the current/working electrode potential relationship in voltammetry are discussed and the essential features of the more important voltammetric methods are presented.

3.1. ELECTROLYSIS CELL, OVERPOTENTIAL, MICRO- AND MACRO-ELECTROLYSIS

In this section you are introduced to the use of electrolysis cells in electro-analytical chemistry and a comparison with electrochemical cells is made. The nomenclatures used for the two types of cell are contrasted.

The concept of overpotential is introduced as a measure of the difference between an observed potential and the reversible potential, and the dependence of overpotential on current is established.

The terms microelectrolysis and macroelectrolysis are defined and are related by examples to analytical methods.

3.1.1. Electrolysis Cell

In Part 2.0 you were introduced to the electrochemical cell and its use in potentiometry and you should now be familiar with the concepts of electrode potential, standard electrode potential and reversible electrodes. The electron transfer process at a reversible electrode is very fast and the corresponding electrode potential is attained instantly. Two such electrodes with a suitable electrolyte constitute a cell and produce, by virtue of the chemical change occurring, an emf. This emf has a maximum value under null current conditions. If a finite current is drawn from the cell the electrodes become polarised, the electrode potentials change and the emf of the cell falls.

∏   What then is an electrolysis cell and how does it differ from an electrochemical cell?

As for an electrochemical cell an electrolysis cell will consist of two electrical conductors (the electrodes) immersed in an electrolyte solution. However in an electrolysis cell the energy source is an externally applied dc voltage and the result is a finite current related in magnitude to the amount of chemical change occurring at the electrodes. A typical electrolysis cell is shown in Fig. 3.1a (see Section 3.2 for a further discussion). The electrodes chosen are inert, ie their chemical composition does not change. They act as electrical conductors and provide a surface upon which the electrode reaction occurs. (See Section 3.3 for further details on choice of electrode materials).

There is also an important difference in the terminology used for electrochemical cells and electrolysis cells. In an electrochemical cell the electrode at which oxidation occurs is termed the negative electrode. We see in Fig. 3.1a that the electrode connected to the positive terminal of the external voltage supply is termed an anode and oxidation occurs at the anode. We have then positive and negative electrodes for reversible cells and anodes and cathodes for electrolysis cells.

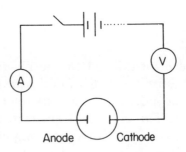

Fig. 3.1a.   *Electroysis cell*

Π   List some differences between an electrochemical cell and an electrolysis cell.

You should have included in your list.

(*a*) The chemical nature of the electrodes in an electrochemical cell usually changes, ie the electrode participates chemically in the cell reaction. This is not so in electrolysis.

(*b*) Electrochemical cells are used under essentially reversible conditions, ie null current conditions. In an electrolysis cell a continuous, changing current is passed.

(*c*) The energy source for an electrochemical cell is the spontaneous chemical reaction occurring in the cell producing an emf. In an electrolysis cell the energy source is an externally applied voltage which imposes a chemical reaction in the cell.

(*d*) The terminology differs; negative (oxidation) and positive (reduction) electrodes for electrochemical cells, anodes (oxidation) and cathodes (reduction) for electrolysis cells.

There are other differences which may be added to this list as we progress through this section.

3.1.2. Overpotential

In order to sustain a finite current in an electrolysis cell the rates of the two electrode reactions and the rate of transfer of ions to the electrodes must all be high. The electrode reaction is always, overall, an electron transfer process but it may be a simple process (one stage) or complex (two or more stages involving adsorption and/or surface reactions). The rate of reaction will be determined by the activation energy barrier to the electrode reaction.

Consider the reaction:

$$0.5\,H_2 + H_2O \longrightarrow H_3O^+ + e$$

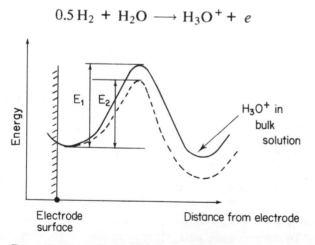

Fig. 3.1b. *Reaction profile for an anodic reaction _____ no applied potential, - - - - potential applied*

an oxidation process occurring at an anode (Fig 3.1b). The two curves in Fig. 3.1b (reaction profiles) represent the effect of a change of electrode potential on the activation energy of the reaction ($E_1 \longrightarrow E_2$). We see that the energies of both the transition state and the hydronium ions are affected, the activation energy is reduced, and hence the rate of reaction is increased.

The additional potential above the reversible electrode potential, that is required to sustain the electrode reaction at a certain rate (cell current at a certain level) is termed the *overpotential*, symbol η. When the overpotential has its origins in the activation energy barrier to electron transfer we have *activation overpotential*. We have then:

$$\eta = \begin{array}{l}\text{measured or applied} \\ \text{electrode potential}\end{array} - \begin{array}{l}\text{reversible or} \\ \text{equilibrium} \\ \text{electrode potential}\end{array}$$

The latter, reversible electrode potential, is E, as used in electrochemical cells.

We may use tables of standard electrode potentials ($E^{\ominus}$), corrected by the Nernst equation to give values of E, as a guide to the potential at which electrolysis reactions will occur. However we must be aware of the possible occurrence of overpotential effects. The activation overpotential is notably high for the evolution of gases particularly on certain electrodes, eg Hg.

There is one other important type of overpotential, namely *concentration overpotential*. This has as its origin the change in analyte concentration occurring in the proximity of the electrode surface (the double layer) due to the electrode reaction.

In an unstirred solution a concentration gradient rapidly develops between the electrode surface and the bulk electrolyte (Fig. 3.1c). An additional potential is therefore required to overcome this gradient and in an extreme case (fast reaction, high current) a limit is reached determined by the maximum rate of transfer of ions to the electrode. Under these conditions a limiting current is reached independent of further changes in potential. This type of overpotential may be removed by stirring the solution. Both types of overpotential (activation and concentration) increase as the current increases.

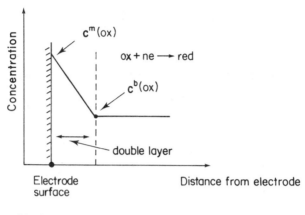

$c^m(ox)$ = Concentration at electrode

$c^b(ox)$ = Concentration in bulk electrolyte

Fig. 3.1c. *Concentration gradient for a cathodic reaction*

The theories which explain the relationship between potential and current are inexact except under certain conditions, and form the subject of electrode kinetics or electrodics. This material is inappropriate to this unit but the approximate shapes of the overpotential/current (η/I) curves are shown in Fig. 3.1d.

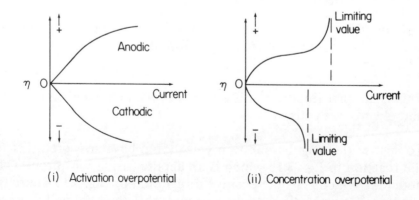

Fig. 3.1d. *Overpotential as a function of current*

∏   For the reactions depicted in Figs. 3.1b and 3.1c in which
       direction will the potential change in order to promote the
       required reaction?

In general the presence of an overpotential of whatever origin will
cause the cell voltage to increase in order to maintain a certain level
of current. Thus (Fig. 3.1e) we have in (*i*) an illustration of the way
the anode and cathode potentials change as the current increases.
Compare this with the effect of overpotential on an electrochemical
cell (*ii*).

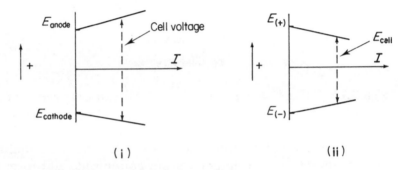

(i)   (ii)

Fig. 3.1e.   *Overpotential in electrolysis cells and
electrochemical cells*

In summary for an electrolysis cell the anode potential increases,
the cathode potential decreases, the cell voltage increases.

For an electrochemical cell the positive electrode potential decreases, the negative electrode potential increases and the cell emf
falls.

This is another important difference between electrochemical cells
and electrolysis cells,

To answer the original question you should have answered that for
the reaction in Fig. 3.1b which is an anodic (oxidation) process the
potential will change in a positive direction. For Fig. 3.1c where the
reaction is a cathodic reduction process the potential will increase
in a negative direction.

One final type of overpotential must be mentioned since it is present in all systems. The origin of this overpotential, namely the *ohmic overpotential*, is in the finite resistance (R) of the electrolyte to the current flow. The relationship between current and this type of overpotential is given by Ohm's law.

$$\eta_{ohmic} = I R \qquad (3.1a)$$

The ohmic overpotential is usually negligible in aqueous electrolyte solutions but must be considered when non-aqueous solutions are used (Section 3.3). It is normally regarded as a property of the cell rather than that of an individual electrode. An exception to this is if a thin film of non-conducting material is deposited on an electrode. In this case the resistance of the film must be included in the ohmic overpotential.

3.1.3. Micro- and Macro-electrolysis

Electrolysis – based methods of analysis rely almost exclusively on the measurement of the relationships between the current in the cell and the potential at a chosen electrode (Section 2.2). The potential is a characteristic of the reaction occurring and forms the basis for qualitative analysis. The current is a measure of the rate of the reaction occurring and forms the basis for quantitative analysis.

Electrolysis – based analytical methods may be divided into two categories, the microelectrolysis methods and the macroelectrolysis methods. If during the course of an analytical experiment the amount of chemical change is small then we have a microelectrolysis method, eg polarography, where currents of a few μA are used for about 1 minute.

∏ The electrolysis of 20 cm^3 of a 10^{-3} mol dm^{-3} aqueous solution of $CuSO_4$ is carried out at an average cell current of 10 μA for 1 minute. Use Faraday's laws to calculate the final concentration of $CuSO_4$ in the solution.

You were introduced to Faraday's laws in 1.4.3. If necessary go back *now* and revise that section. In this experiment, quantity of electricity passed, $It = 10 \times 10^{-6} \times 1 \times 60 = 6 \times 10^{-4}$ C.

∴   Number of Faradays passed = quantity of electricity/F = 6.2 $\times 10^{-9}$.

∴   Number of mol of Cu^{2+} deposited on cathode = 3.1×10^{-9}.

Number of mol of Cu^{2+} in 20 cm^3 of 10^{-3} mol dm^{-3} solution

$$= \frac{20}{1000} \times 10^{-3}$$
$$= 2 \times 10^{-5}$$

Number of mol of Cu^{2+} remaining after electrolysis

$$= (2 \times 10^{-5} - 3.1 \times 10^{-9}) \simeq 2 \times 10^{-5},$$

ie about 0.01% decrease.

Hence the concentration is effectively unchanged,

This is not an unrealistic conclusion. Currents in polarography are rarely as high as 100 μA, and electrolysis rarely exceeds 10 min

∏   What would your conclusion be if 100 μA were passed for 10 min in the above experiment?

The final concentration is still about 2×10^{-5} mol dm^{-3} but this time there would have been about a 2% decrease in concentration.

We see that repeated analyses may be carried out on the same solution without appreciable change in the analyte concentration. This property can be extremely important, as one often faces the situation of limited availability of analyte and the need to use this for as many different tests as possible.

You should have learned already of the classification of all analytical methods as being either destructive or non-destructive.

Π   Name one destructive and one non-destructive method of analysis.

The answer to this will depend upon your experience but a good example of a destructive technique is atomic absorption spectroscopy (AAS). The analyte here is burned in a flame and vented to the atmosphere.

Good examples of non-destructive techniques are infrared and ultraviolet spectroscopy.

Try to place all the analytical techniques that you have experienced into these categories. You will find some grey areas, eg gravimetric analysis. Here the sample loses its identity by virtue of the precipitating reaction but is often recoverable, if necessary, by decomposing the precipitate.

Microelectrolysis methods are then almost completely non-destructive.

If during the course of an analytical experiment the amount of chemical change is large then we have a macroelectrolysis method, eg *coulometry*, where currents of a few mA are used for ten minutes to one hour. What constitutes a large change? If the concentration of the analyte changes more than 10% we may say macroelectrolysis has occurred. You might suppose there could exist a gradual change from microelectrolysis methods to macroelectrolysis methods. In practice methods fall naturally into those where very little electrolysis occurs and those where almost all of the analyte is consumed.

Π   The electrolysis of 20 cm^3 of a 10^{-3} mol dm^{-3} aqueous solution $CuSO_4$ is now carried out for 30 minutes at an average cell current of 2 mA. Use Faraday's laws to calculate the final concentration of $CuSO_4$ in the solution.

This is exactly the same as the previous question and the supplementary to that question in the sense of the method of solution.

However this time the quantity of electricity passed $= 30 \times 60 \times 2 \times 10^{-3} = 3.6$ C.

$\therefore$ mol Cu^{2+} deposited $= 1.86 \times 10^{-5}$

After electrolysis there remain $(2\text{--}1.86) \times 10^{-5}$

$$= 0.14 \times 10^{-5} \text{ mol } Cu^{2+}$$

$\therefore$ Concentration is now $0.14 \times 10^{-5} \times \dfrac{1000}{20}$

$$= 7 \times 10^{-5} \text{ mol dm}^{-3}$$

There has been a 93% reduction in the concentration of $CuSO_4$. Clearly macroelectrolysis.

Macroelectrolysis methods are also usually non-destructive since the analyte often ends up plated on an electrode and can be recovered.

SAQ 3.1a Explain the differences in operation and terminology between an electrochemical cell and an electrolysis cell.

SAQ 3.1b
> Define overpotential at an electrode. How will the potential of a reduction reaction occurring at a cathode change as the current in the cell is increased?

SAQ 3.1c
> Give one example of a technique falling into each of the categories: microelectrolysis, macro-electrolysis, destructive, non-destructive.

SUMMARY AND OBJECTIVES

Summary

You have been introduced to the use of electrolysis cells in electro-analytical chemistry. A comparison with the already familiar subject of electrochemical cells is made and the important differences in terminology between electrolysis cells and electrochemical cells established.

The concept of overpotential is introduced and related to the electrode potentials at which electrode reactions occur and to the current flowing in the cell.

Electrolysis based electro-analytical methods are categorised into microelectrolytic and macroelectrolytic methods.

Objectives

You should now be able to:

- distinguish between the characteristics of an electrochemical cell and an electrolytic cell;

- explain the terminologies negative and positive for electrochemical cells, anode and cathode for electrolytic cells;

- define overpotential; list three types of overpotential and explain their origins; and explain the effect of overpotential on the observed potential at which an electrode reaction will occur;

- distinguish between micro-and macro-electrolytic methods and give examples of each type.

3.2. THREE ELECTRODE CIRCUITRY, THE POTENTIOSTAT AND CELL DESIGN

This section deals first with the principles of 3-electrode circuitry and compares them favourably with those of 2-electrode circuitry. The functions of the working electrode (WE), secondary electrode (SE) and reference electrode (RE) in a 3-electrode circuit are shown and the principles of potentiostatic control of the potential of the working electrode are explained.

The features of a well-designed electrolysis cell are listed and considered particularly in the context of a cell to be used for voltammetric methods of analysis.

3.2.1. The Problems Associated with Two Electrode Circuitry

Consider the circuit (Fig. 3.2a). In all electrolysis - based analytical methods we are interested in the reaction at one electrode only. This electrode is termed the working electrode (WE) and it may be either an anode or a cathode depending on the nature of the analytical reaction. For example an analytical method based on the reduction of copper(II) ions, $Cu^{2+} + 2e \longrightarrow Cu$, would have a cathode as the WE since reduction is occurring.

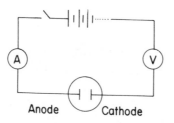

Anode   Cathode

Fig. 3.2a. *2-electrode circuit*

In a two-electrode circuit, the electrode other than the WE, is termed the secondary electrode, (SE), (alternative names are auxiliary electrode and counter electrode). There is a tendency to ignore the chemistry occurring at the SE. This can sometimes cause problems and you will be reminded from time to time of the need to consider the SE reaction and to allow for it in the cell design. The voltmeter reading in the circuit of Fig. 3.1a is a measure of the difference between the electrode potentials of the two electrodes together with the ohmic drop (IR) due to the resistance of the electrolyte (R) between the electrodes.

$$V = E(\text{anode}) - E(\text{cathode}) + IR \qquad (3.2a)$$

where V is the voltmeter reading, ie cell voltage.

∏   If the cell current in a two electrode cell is 10 μA and the cell resistance is 20 Ω what are the potentials of the anode and cathode when the voltmeter reads 2 V?

There is no answer to this question and that in itself makes an essential point, or if you prefer, the answer is that it is impossible to calculate $E(\text{anode})$ and $E(\text{cathode})$. You may calculate

$$IR = 10 \times 10^{-6} \times 20 = 2 \times 10^{-4} V$$
$$\therefore \quad V = E(\text{anode}) - E(\text{cathode}) + 2 \times 10^{-4}$$

Hence $E(\text{anode}) - E(\text{cathode}) = 2 - 2 \times 10^{-4} \simeq 2V$

However you cannot then separate the contributions of $E(\text{cathode})$ and $E(\text{anode})$.

Many methods use aqueous solutions of strong electrolytes and in these cases the ohmic drop is negligible. If we ignore this factor then:

$$V = E(\text{anode}) - E(\text{cathode}) \qquad (3.2b)$$

but the problem of separating the anodic and cathodic contributions still remains. If one of the electrodes has a potential which remains effectively constant as the cell voltage and current vary then it may be regarded as a pseudo-reference electrode. Suppose the WE is a cathode, then:

$$V = E_{SE} - E_{WE} \qquad (3.2c)$$

If E_{SE} remains effectively constant we may state a value for E_{WE} with respect to the potential of the SE as a reference.

∏   If the cell voltage is 2 V and the ohmic drop is negligible state the potential of a cathode WE with respect to the SE whose potential may be assumed constant.

The answer is $E_{WE} = -2\ V$ with respect to the potential of the secondary electrode.

$$V = E(\text{anode}) - E(\text{cathode}) + IR$$
$$= E_{SE} - E_{WE} + 0$$
$$\therefore \quad E_{WE} - E_{SE} = -2\ V$$

Similarly if the cell voltage changes from V_1 to V_2 and the ohmic drop is negligible and $E(\text{anode})$ (SE) is constant then:

$$\Delta V = V_2 - V_1$$
$$= (E_{2,SE} - E_{1,SE}) - (E_{2,WE} - E_{1,WE})$$
$$= 0 - \Delta E_{WE}$$
$$= - \Delta E_{WE} \qquad (3.2d)$$

Thus we have an important result that a change in the measured cell voltage reflects a change in E_{WE}.

Two electrode circuitry prevailed in electro-analytical chemistry into the late 1950's, the best known example being dc polarography (Section 3.5). The integrity of the method was dependent upon the potential of the secondary electrode remaining constant over a range of cell current – an unsatisfactory situation.

If non-aqueous solvents are used then one can sometimes no longer ignore the ohmic drop and this factor (IR) will vary as the cell voltage and hence cell current vary. In this situation even with E_{SE} a constant, E_{WE} is unknown.

∏   Calculate the resistance of a 2 cm length of 0.1 mol dm^{-3}
       aqueous KCl between two electrodes, each of area 2 × 10^{-5}
       m^2. Conductivity of 0.1 mol dm^{-3} aqueous KCl is 1.29 S m^{-1}

Conductance = $G = \kappa/J$

where κ/S m^{-1} is the conductivity and J the cell constant, defined
as L/A where L is the separation between the electrodes (m) and A
the area of cross-section of the electrodes (m^2)

$$G = \frac{1.29 \times 2 \times 10^{-5}}{0.02} = 1.29 \times 10^{-3} S$$

Thus $R = 1/G = 775\Omega$

We see that for a typical microelectrolysis method with the cell current about 5 μA, the ohmic drop would be 5 × 10^{-6} × 775 × 10^3
= 3.9 mV.

It would be an improvement if a method could be found that enables
us to measure and control the value of E_{WE} unambiguously.

3.2.2.   Three Electrode Circuitry

The solution to the problems described in the previous section are
largely overcome by the introduction of a third electrode into the
circuit (Fig. 3.2b).

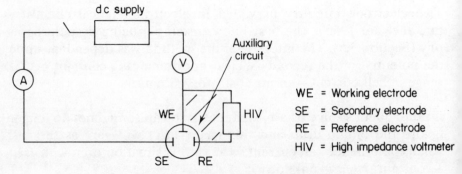

Fig. 3.2b.   *3-electrode circuit*

An auxiliary circuit, often termed a potentiometer circuit, has been added to the previous two-electrode circuit. The cell current still passes between the SE and the WE in the primary circuit and the cell voltage (V) is still developed between these electrodes. We still have:

$$V = |E_{WE} - E_{SE}| + IR \qquad (3.2e)$$

The modulus sign indicates that we take the positive value of the difference in the two values. This is to allow for the WE being either an anode or a cathode.

However this equation (3.2e) is now irrelevant since we are interested only in E_{WE} and E_{SE} is not constant in these circuits. In many modern instruments the cell voltage is not displayed.

Π   Why is the equation irrelevant and why is the cell voltage redundant?

Consider the auxiliary circuit comprising the WE, the reference electrode (RE) and a high impedance voltmeter (HIV). If P is the potential difference measured by the HIV then:

$$P = |E_{WE} - E_{RE}| + IR' \qquad (3.2f)$$

where R' is the resistance in the electrolyte path between the WE and the RE. This is a null current potentiometer circuit and not part of the primary current carrying circuit of the cell, this is an essential point to remember.

Provided $IR' \longrightarrow 0$, for the reasons given previously:

$$P = |E_{WE} - E_{RE}| \qquad (3.2g)$$

For a reference electrode we choose an electrode with a known stable potential and a small temperature dependence of potential. The best known of such electrodes is the saturated calomel electrode (2.2.6).

∏   Why did we not use a SCE as an electrode of fixed potential
       in a two-electrode circuit?

A saturated calomel electrode is a reversible electrode and is designed to give a fixed potential under null current conditions. If a
finite current passes through it polarisation occurs and the potential
changes. The electrode returns to its original potential if the current
is removed provided the current flows for only a short time.

We quote the values of the E_{WE} with respect to the potential of the
reference electrode.

Thus substituting E_{SCE} in Eq. 3.2g, gives

$$P = |E_{WE} - E_{SCE}|$$
$$P = E_{WE} (SCE) \qquad\qquad (3.2h)$$

If the measured voltage on the high impedance voltmeter is -0.600
V, then $E_{WE} = -0.600\ V\ (SCE)$

Where does the minus sign come from? The voltmeter will either
display both positive or negative values or the operator has to note
the polarity of the connections made. Values of E_{WE}, obtained with
reference electrodes other than SCE, are usually converted to the
SCE scale before reporting the results.

∏   Convert the following potentials to the SCE scale, all at 25 °
       C:

       (*a*) $E_{WE} = +0.012\ V$ (saturated aq Ag, AgCl/Cl$^-$);

       (*b*) $E_{WE} = -0.230\ V$ (SHE).

You will have needed to consult the tables of standard electrode
potentials given in Figs. 2.2c and 2.3b

If you did not realise this, consult the tables and try the question
again.

$$E^{\ominus}(H^+, H_2, Pt) \quad = 0 \ V$$

$$E^{\ominus}(Cl^-, AgCl, Ag) \ = 0.222 \ V$$

$$E^{\ominus}(cal) = 0.241 \ V$$

$$\therefore \quad (a) \ E_{WE} = + 0.012 + 0.222 - 0.241$$

$$= 0.007 \ V \ (SCE)$$

$$(b) \ E_{WE} = - 0.230 - 0.241$$

$$= - 0.471 \ V \ (SCE)$$

We now have a reliable measure of E_{WE}: Also,

$$\Delta P = \Delta E_{WE} \qquad (3.2i)$$

so we can monitor changes in E_{WE}. The only uncertainty arises when solvent systems of a high resistance are used making the ohmic drop (IR') no longer negligible. This is usually due to the introduction of non-aqueous solvents and these can also cause the reference electrode to become unstable.

It remains to devise a method to automatically select, change and monitor E_{WE} in a controlled manner

3.2.3. Potentiostatic Control

Since the early 1960's commercial apparatus designed for electrolysis – based analytical methods has had as its brain an electronic device known as a potentiostat (not to be confused with a potentiometer). In order to fully understand the mode of operation of a potentiostat some knowledge of electronics is required. This knowledge must extend at least to an understanding of operational amplifiers. Those of you with this knowledge should refer to the textbooks given as references where you will find circuit diagrams and further explanation.

Fortunately, it is not necessary to have this depth of understanding to appreciate the function of a potentiostat. The function of a potentiostat is to control the potential of the WE with respect to E_{RE}. The potentiostat enables one to hold this potential constant or to vary the potential in a controlled manner to a pre-selected pattern. To do this a three electrode circuit is necessary (Fig. 3.2b). The required E_{WE} (SCE), is fed into the potentiostat control and when the circuit is completed the measured HIV reading is:

$$P = E_{WE} \text{ (SCE) (measured)}$$

and this value is fed instantly to the potentiostat.

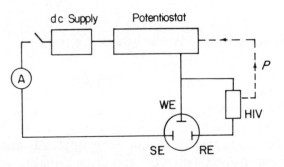

P = potential difference $E_{WE} - E_{RE}$

Fig. 3.2c. *Potentiostatic control*

The potentiostat compares the measured value of E_{WE} with the required value of E_{WE} and any difference is an error signal. If the error signal is zero the system is under potentiostatic control. If the error signal is finite the potentiostat causes a change in the dc power to the cell in such a direction as to decrease the error signal. This response is very rapid and the system comes under potentiostatic control within microseconds.

An extension of this is to have available the capability to scan a range of potential in a pre-determined manner. An initial E_{WE} is fed into the potentiostat control together with a final E_{WE} and the required scan pattern.

For example

$$\text{initial } E_{WE} = -0.10 \ V \ (SCE),$$
$$\text{final } E_{WE} = -0.90 \ V \ (SCE),$$
$$\text{linear potential scan} = 5 \ mV \ s^{-1}.$$

In this case when the circuit is completed the initial E_{WE} is rapidly established as described above. The scan commences and the potentiostat continuously monitors and adjusts E_{WE} to conform to the selected pattern.

Π   For the linear scan pattern, given in the example above, calculate the reading on the HIV that will cause a zero error signal 2 min after the scan commences.

The answer is −0.70V.

A zero error signal occurs when:

$$E_{WE} \text{ (measured)} = E_{WE} \text{ (required)}.$$

The required E_{WE} is given by:

$$E_{WE} = -0.10 - (5 \times 10^{-3} \times 2 \times 60)$$
$$= -0.70 \ V \ (SCE)$$

$$\therefore \quad P = E_{WE} \text{ (measured)} = -0.70 \ V \ (SCE)$$

It is worth reminding you here that the high impedance voltmeter will either read −0.70 *V* because it can display both positive and negative values, or it will read 0.70 *V* and you will see by the polarity of the connections that the WE is negative compared with the RE. A final possibility if it can only display positive values is that you have connected it with the wrong polarity and no reading will be obtained until you reverse the connections.

Some more sophisticated versions of potentiostatically controlled equipment have the facility to select patterns of potential change other than linear. The pattern most frequently encountered is the cyclic or triangular pattern; eg cyclic voltammetry (Section 3.4 and Fig. 3.2d).

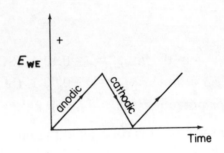

Fig. 3.2d. *Cyclic linear voltage sweep*

Potentiostatic control as described above finds widespread use in electro-analytical chemistry. Far less frequently it is necessary to control the cell current. Circuits similar to that in Fig. 3.2c; ie modified potentiostat circuits, or specially designed different electronic circuits are available. Such a device, ie a circuit for controlling the current is termed a galvanostat.

3.2.4. Cell Design

Basically we require a three electrode cell Fig. 3.2e.

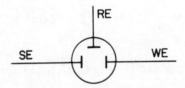

Fig. 3.2e. *Schematic design of 3-electrode cell*

Usually there are other requirements, commonly provision for some or all of the following, depending upon the analytical technique:

(*a*) stirring of the solution;
(*b*) thermostatting;
(*c*) emptying the cell *in situ*;
(*d*) adding chemicals;
(*e*) purging with gas;
(*f*) gas venting;

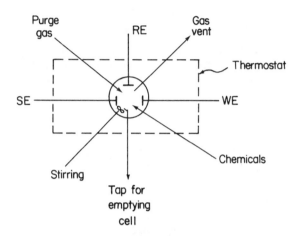

Fig. 3.2f. *Schematic diagram of 3-electrode cell with additional requirements*

A well designed cell will provide as many of these features as necessary together with minimisation of the cell volume. A suitable cell for microelectrolysis methods, eg polarography, would be as in Fig. 3.2g, note the pear shape used to minimise the volume. We may pick out two features for a more detailed treatment.

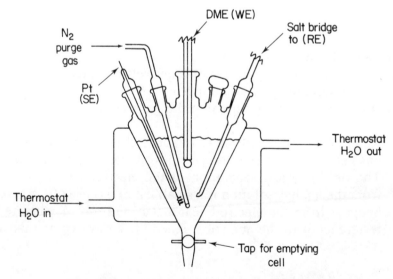

Fig. 3.2g. *Typical microelectrolysis cell*

(*i*)   Device for gas purging. The purging gas, usually nitrogen, may
        be introduced via a two-way tap in a device depicted in Fig.
        3.2h. This design is given as an example. In one position of the
        tap gas flows down the tube and through a frit into the base of
        the analyte solution. In the second position the gas enters the
        neck of the cell and blankets the surface of the analyte solu-
        tion thus preventing re-entry of air into the solution. Cylinder
        nitrogen (white spot) contains a few ppm of oxygen and if it is
        necessary to further reduce the oxygen level in the solution the
        nitrogen supply must be scrubbed of oxygen.

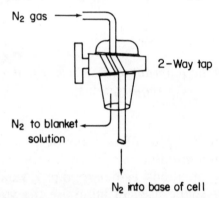

Fig. 3.2h.   *Gas purging and blanketing device (as drawn nitrogen
flows into base of cell)*

One method of doing this is to pass nitrogen through vanadium chloride (II oxidation state, purple). Alternatively commercial cartridges
filled with a scrubbing formulation based on chromium(III)oxide are
available as disposable inserts into the gas stream.

(*ii*)   Reference electrode bridge. In some circumstances the refer-
         ence electrode, eg SCE, may be introduced direct into the cell.
         The only physical separation of the electrode inner solution
         from the analyte solution is a small frit or a piece of glass wool.
         Seepage from the electrode can occur and this may create in-
         terference with the analytical reaction. To overcome this, and

also as part of the usual design practice, the electrode is placed in a separate compartment, and a salt bridge is formed to the analyte solution (Fig. 3.2i)

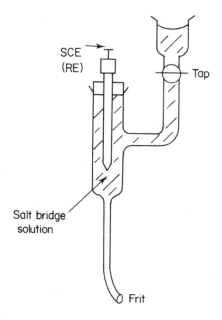

SCE (RE)

Tap

Salt bridge solution

Frit

Fig. 3.2i. *Reference electrode salt bridge*

The salt bridge solution is usually the supporting electrolyte solution used for the analysis, (Section 3.3.3). Mention should be made of a special form of such a reference electrode system where the tip of the salt bridge is bent to a position close to the WE (Fig. 3.2j)

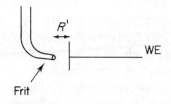

R'

WE

Frit

Fig. 3.2j. *Luggin probe*

This device is called a Luggin probe and its purpose is to minimise the ohmic drop (IR'). This may become important for non-
electrolyte solutions.

Π   Why is it necessary to minimise the ohmic drop (IR')?

In the auxiliary (potentiometer) circuit of the three electrode system
we have seen that,

$$P = |E_{WE} - E_{RE}| + IR' \qquad (3.2f)$$

Only if $IR' \longrightarrow 0$ does

$$P = |E_{WE} - E_{RE}| = E_{WE}\ (SCE)$$

Only if this is true can we take P (the high impedance voltmeter
reading) to be a measure of E_{WE} (SCE). The integrity of potentiostatic control depends upon this.

If a macroelectrolysis method is being used essentially the same
features are required as for microelectrolysis. One additional feature
for macroelectrolysis is provision for the isolation of the secondary
electrode in a compartment separated from the analyte solution by
a porous frit. This is done to prevent products from the secondary
electrode reaction from migrating to the working electrode.

Π   List the features required in the design of a good cell for
       analytical methods based on electrolysis.

This requires a direct production of the information given in the
Section 3.2.4.

(a) 3-electrodes, (b) potentiostatic control, (c), provision for stirring
of the solution, (d) provision for purging the solution of O_2 and
preventing re-entry of air, (e) provision for emptying cell *in situ*,
(f) provision for adding chemicals, (g) suitable vent to allow gases
to escape, (h) provision for thermostatting, (i) suitably designed RE
bridge, and (j) for macroelectrolysis, provision of a SE compartment
separated by a frit from the analyte solution.

All of these are good design features but not necessarily all are required for any one method.

Alongside these features the volume of the cell should be kept to a minimum.

SAQ 3.2a

> Sketch and label a circuit for a potentiostatically controlled three electrode cell and explain how potentiostatic control is achieved and maintained. Why are three electrodes necessary? What problems arise when non-aqueous solutions are used?

SAQ 3.2b

> Sketch a cell suitable for microelectrolysis and comment on the features. In what way would you change the design in order to carry out a typical macroelectrolysis.

SAQ 3.2c   Consider the following three electrolytes;

(*i*) H_2O, (*ii*) 0.1 mol dm^{-3} aqueous KCl, (*iii*) 1.0 mol dm^{-3} $(C_2H_5)_4NBF_4$ in CH_3CN,

with conductivities; (*i*) $\kappa = 10/^{-4}$ S m^{-1}, (*ii*) $\kappa = 1.289$ S m^{-1}, (*iii*) $\kappa = 5.55$ S m^{-1}.

If two electrodes of equal area (0.2 cm^2) are placed in these solutions when a cell current of 20 μA is passing calculate the distance apart of the electrodes if an ohmic drop of <1 mV is to be achieved. Comment on the implications of your results in electro-analytical applications.

SUMMARY AND OBJECTIVES

Summary

A criticism of 2-electrode circuitry is presented based on the uncertainty of the value of E_{WE}. The use of three electrodes removes this uncertainty and produces an unambiguous value for E_{WE} (SCE).

The use of a potentiostat to control the value of E_{WE} (SCE) is described. Finally the necessary components of an electrolysis cell are listed and considered from the point of view of good cell design, particularly for microelectrolysis methods.

Objectives

You should now be able to:

● criticise the use of 2-electrode circuitry for electrolysis methods;

● draw a labelled circuit diagram for use with a 3-electrode cell and explain the function of the auxiliary (potentiometer) circuit;

● define WE, SE, and RE;

● explain the function of a potentiostat and draw a labelled circuit diagram showing the use of a potentiostat in controlling the WE potential;

● state the function of a galvanostat;

● summarise the essential features of a well designed electrolysis cell and describe the function of a Luggin probe.

3.3. THE ELECTRODES, SOLVENT AND SUPPORTING ELECTROLYTE SYSTEM

We have seen that three electrodes are used in modern electrolysis-based analytical methods. One of these electrodes, the reference electrode (RE), does not participate in the primary current carrying circuit and so need not be considered further here. Sufficient has been said about reference electrodes in Sections 2.2 and 3.2.

∏   List the ideal properties of a good reference electrode and name two such electrodes.

You should have included in your list:

- stable potential;
- reversible, rapid Nernstian response;
- minimal Nernstian effect due to concentration changes;
- rapid return to reversible potential after a small current has passed (some polarisation);
- no drift in potential when a small current is passed continuously;
- stable chemically to the solutions in which it is used;
- a small change in potential with change in temperature;

Two such electrodes are the saturated calomel electrode (SCE) and the saturated silver–silver chloride electrode. The latter is preferred in non-aqueous solutions because it gives a more stable potential than the SCE in these solutions. Revise Section 2.2 if you did not obtain most of the above properties.

We shall consider here the choice of material for use as working and secondary electrodes. These two electrodes together with the solvent and supporting electrolyte comprise a system which would be regarded as a package designed by you, the analyst, to ensure an optimum mix of the following features.

(*a*) A stable working electrode at which the required electrochemical reaction can take place efficiently.

(*b*) A stable secondary electrode capable of sustaining a satisfactory cell reaction.

(*c*) A homogeneous solution, ie all components are mutually soluble at all stages in the analytical procedure.

(*d*) A solution with sufficiently low electrical resistance to ensure reliable control of the electrochemical variables, principally the potential of the working electrode.

(*e*) A voltage window with anodic and cathodic voltage limits as wide as possible.

All of these features are strongly inter-related and we shall consider these relationships after dealing separately with the electrodes, the solvent and the supporting electrolyte. By that time the meaning of a voltage window will be clear.

⊓   In items (*a*) and (*b*) above what do you think is meant by the 'required' electrochemical reaction, and a 'satisfactory' cell reaction?

At an electrode it is possible for more than one reaction to be occurring simultaneously. The 'required' electrochemical reaction is the reaction in which the analyte is either oxidised or reduced at an electrode producing the current which is a measure of the analyte concentration. If the required reaction is the only reaction occurring the faradaic efficiency for the electrode process is 100%, for example in the analysis for Cu^{2+} by reduction at a Hg cathode the required reaction is

$$Cu^{2+} + 2e \longrightarrow Cu(Hg)$$

A reaction is also occurring at the SE, to the same extent as that at the WE. This reaction at the SE should not alter in any way, by reason of the reaction products, the nature of the analyte or the reaction at the WE. This interference by the SE reaction can happen if reaction products diffuse or migrate to the WE. A satisfactory cell reaction is one in which no interference occurs from the SE reaction and the required reaction occurs at the WE with 100% Faradaic efficiency.

⊓   Why is a low electrical resistance important for reliable control of the E_{WE}?

You learnt (Section 3.2) that the key to the control of the WE potential is the auxiliary circuit connecting the WE and the RE. The ohmic drop (IR*l*) between the WE and RE must be as low as possible to achieve reliable control. Revise Section 3.2 if you are unsure on this point.

In addition to the above features which may be regarded as essential, it is desirable for the material used to be easy to handle, available pure or easily purified, non-toxic, and of low cost.

3.3.1. Working and Secondary Electrodes

The materials for the working and secondary electrodes should be inert in the sense that they do not undergo any change during the electrode reactions. They are required to provide an electrical conductor and a surface upon which the required reactions can occur efficiently. These reactions may involve the analyte in a variety of processes including adsorption on the electrode, surface reactions, and electron transfer.

∏   Which property of the electrode will measure the efficiency of the electrode process?

If the reaction occurs efficiently the activation energy is small and the reaction will have a low overpotential. Thus a good electrode will cause the overpotential to be a minimum and the potential for the reaction will be close to the reversible electrode potential. You will see later that a low overpotential is not always desirable for non-required reactions (see later in this section). For the required analyte reaction a low overpotential is desirable and this is a feature of metallic cation deposition reactions.

In this unit it is not appropriate to discuss the relative merits of the shapes and sizes of electrodes or the variety of analyte reactions that may occur at the electrodes. These details will be discussed in the context of the individual techniques. The electrode materials most often used are mercury, platinum and carbon. Before discussing these materials in turn let us consider a reaction which commands priority attention in determining electrode/solvent stability. Any solvent containing water also contains hydronium (H_3O^+) ions. Solvents containing water are very common in electro-analytical chemistry so the cathodic reduction of the hydronium ion is of vital interest. We may write the reaction as

$$H_3O^+ + e \xrightarrow{M} 0.5\,H_2 + H_2O \qquad (3.3a)$$

where M is Hg, Pt, C.

From a practical point of view the potential of the working electrode at which a measurable current begins to develop, as a consequence of some process other than the desired analyte reaction, defines the voltage limit for the system. The unwanted reaction will interfere with any required analyte reaction that occurs at a potential beyond this limit. The hydronium ion (H_3O^+) reduction reaction often sets the cathodic voltage limit of the system.

Three factors determine the cathodic voltage limit set by the above reaction.

(*a*) The reversible electrode potential of the reaction, $E^\ominus = -0.241$ V (SCE) at 25 °C, and identical for mercury, platinum and carbon.

(*b*) The overpotential of the reaction. This varies with the nature of the electrode material.

(*c*) The concentration of the hydronium ions in the solution. This effect is common to all three electrode materials and is illustrated in Fig. 3.3a.

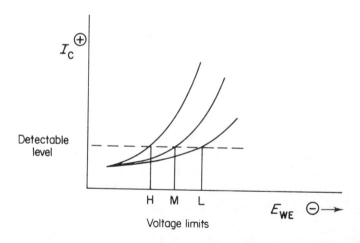

Fig. 3.3a. *Voltage limit as a function of concentration of electro active species. H, M, L refer to high, medium and low concentration of H_3O^+ respectively*

We see that the voltage limit becomes progressively less negative as the concentration of hydronium ions increase.

∏   From the following data, which of the materials, mercury, platinum or carbon is preferred as a cathode WE?

Overpotential values for the reaction $H_3O^+ + e \longrightarrow 0.5\,H_2 + H_2O$ at 25 °C in 0.1 mol dm^{-3} aqueous HCl are as follows:

Hg, -1.06 V;   Pt, -0.02 V;   C, -0.78 V.

Clearly here Hg is preferred because it has the highest cathodic voltage limit. Pt would be a poor choice.

Let us now consider the three electrode materials in turn. We shall assume that the types of analyte in mind are inorganic anions and cations unless otherwise stated.

(*i*)   Mercury is very toxic and demands a neat and tidy working practice. One of the first electro-analytical techniques that you are likely to encounter either in industry or in a teaching laboratory is some form of polarography.

Mercury is a liquid under normal experimental conditions and thus can be used in a variety of configurations. For example, as drops on the end of a capillary, or as a thin film coated on other materials. In all cases mercury presents a pure, smooth, and homogeneous surface.

An important property of mercury is that the potential at which the hydronium ion reduction reaction occurs is much more negative than for other electrode materials, eg about -1.0 V (SCE) in 1 mol dm^{-3} aq $HClO_4$. This reaction on mercury sets the cathodic voltage limit in aqueous solutions unless the supporting electrolyte cations are more easily reduced (3.3.3).

On the anodic side, mercury is a poor choice of electrode material because it is readily oxidised, particularly in the presence of certain anions. Anions such as Cl^-, CN^-, OH^- react with mercury to give either soluble complex anions or precipitates. These reactions occur at anode potentials in the range -0.1 to $+0.2$ V (SCE) and normally impose the anodic voltage limit for mercury electrodes.

Thus mercury is the material of choice for working electrodes where the electrode process is cathodic (reduction), but mercury is rarely used as a working electrode for anodic processes.

(*ii*) Platinum is a well established solid electrode material. It can be obtained in a high state of purity as sheet, rod, wire or gauze, can be sealed into glass fairly easily, and is non-toxic.

It is however far less useful than mercury as a cathode material because the reduction reaction of the hydronium ion on a platinum surface occurs at a much lower negative potential, eg about -0.2 V (SCE) in 1 mol dm^{-3} aq $HClO_4$.

The anodic voltage limit is set by the decomposition of the solution, usually at a potential more positive than $+1$ V (SCE). This excellent anodic behaviour makes platinum the material of choice as the secondary electrode for methods where the working electrode is a cathode. Platinum is also a good choice as the working electrode for methods based on an anodic analyte reaction.

∏ Why has platinum a less negative cathodic voltage limit than has mercury for the reduction of the hydronium ion?

We see that the difference is between -0.2 V (SCE) on Pt and -1.0 V (SCE) on Hg. This considerable difference is due to the fact that Pt is about 10^9 times more effective as a catalyst for the reaction than Hg. This causes the overpotential for the reaction on Pt to be much less than on Hg.

This is really asking the question 'which property of an electrode will measure the efficiency of the electrode process' (3.3.1) in a different way.

(*iii*) Carbon. The two best known types of carbon used as electrode materials are glassy carbon (vitreous) and wax impregnated graphite. The latter is the older of the two types and consists of porous graphite impregnated under vacuum with wax, and usually used in rod form. Wax impregnation prevents penetration of gases and liquids into the pores and gives a uniform surface with stable electrical properties. Glassy carbon is available as rod, flat disc or plate and can be polished to a mirror finish, again providing a uniform surface with reproducible electrical properties. Glassy carbon is readily fabricated into a variety of electrode configurations.

Carbon falls between mercury and platinum in terms of the cathodic voltage limit set by the reduction of hydronium ions, about -0.9 V (SCE) in 1 mol dm^{-3} aq HClO$_4$. As an anode material it is comparable with platinum, and the anodic voltage limit for carbon, as for platinum, is set by the decomposition of the solution.

Carbon is rarely used directly as a working electrode for inorganic analysis. However it is easy to plate wax-impregnated graphite with a thin film of mercury. The electrode then takes on the properties of a mercury electrode (in rod form) and as such is the electrode of choice in the technique of anodic stripping voltammetry (3.5.1g).

∏   Suggest a method for plating carbon with a thin film of mercury?

This may be done by electrolysing a dilute aqueous solution of a soluble mercury(II) salt, eg mercury(II)nitrate, using a wax-impregnated carbon electrode as the cathode. The conditions would conform to those of the technique, coulometry at controlled potential (4.2).

Do not be concerned if this question puzzled you. The information required was not really in the text. However if you are beginning to understand electrolysis the above response would have been a reasonable deduction.

The most important property of carbon as an electrode material is that it forms a good catalytic surface for organic reactions, better than platinum and much better than mercury. Hence the main use of carbon electrodes has been in organic analysis and in the study of organic reaction mechanisms, and these are rapidly developing fields of applications. Glassy carbon has been used extensively as the working electrode in electrochemical detectors used as part of high performance liquid chromatography (hplc) apparatus.

∏   Summarise your conclusions about the choice between mercury, carbon and platinum as electrode materials for use as working electrodes.

In general the choice of electrode material for inorganic analysis would be:

| Cathodic | WE | $Hg > C \gg Pt$ |
| Anodic | WE | $Pt \simeq C \gg Hg$ |

Carbon is the preferred material when the analytes are organic.

Over and above these general points the electrode must be capable of supporting the required analyte reaction.

3.3.2. Solvents

The very nature of electrolysis, ie the passage of ionic species through an electrically conducting solution, dictates that conditions exist in solution capable of sustaining stable ionic species. This, of necessity, requires that the solvent has some degree of polar character to be capable of solvating these ions.

∏   What is the physical property of a solvent that is usually used as a measure of polarity?

You should have answered the relative permittivity (or dielectric constant) (Fig. 1.2a) and

Highly polar, H_2O, $\epsilon_r = 78.5$ at 25 °C
Non polar , CCl_4, $\epsilon_r = 2.2$ at 25 °C.

Both pure protic and pure aprotic solvents have been used; mixtures of the two types provide a continuous range of polarity.

∏   Explain with examples what you understand by a protic solvent and an aprotic solvent.

Protic solvents exchange protons rapidly and are strong hydrogen donors. The hydrogen atoms in these solvents are bound to more electronegative atoms.

Aprotic solvents have hydrogen bound to carbon and are very weakly acidic, are poor hydrogen bond donors and exchange protons only very slowly.

(You should read again Section 1.1 which covers the fundamentals of solvent structure and behaviour).

Commonly used protic solvents are water, methanol, ethanol and mixtures of these three. Commonly used aprotic solvents are acetonitrile, CH_3CN, dimethyl sulphoxide, $(CH_3)_2SO$ (DMSO) and dimethylformamide $HCON(CH_3)_2$ (DMF). These aprotic solvents are not usually used in mixtures with each other but are often used in mixtures with protic solvents forming a solvent class sometimes termed polar-aprotic.

Most of the reported analytical work using electrolysis – based methods has been done using H_2O, H_2O-CH_3OH or $H_2O-C_2H_5OH$ with other solvents introduced only to facilitate the solubility of the analyte and/or the supporting electrolyte. This is because in the past metallic cation analysis dominated the field. As interest in the analysis of organic materials has grown in importance and, with the continuing use of electro-analytical methods in organic mechanism studies, the use of aprotic solvents has increased. Most of the development work is now related to the analysis of organic chemicals, notably in the analysis of pharmaceuticals.

It is necessary to have an electrically conducting solution with a low electrical resistance (3.2.1, 3.2.4) For the aprotic solvents the supporting electrolyte plays a vital role in providing the necessary conducting solution. We have seen that the presence of hydronium ions in the solution is a very important consideration, (3.3.1) and any solvent containing water will contain hydronium ions.

A further important property of protic solvents is that they are able to dissolve oxygen to an appreciable extent (about 10^{-3} mol dm^{-3} or 30 ppm in water at 25 °C, 1 atm). Oxygen is electroactive and is reduced (at a cathode) in aqueous solutions in two stages at potentials of about -0.5 V (SCE) and about -0.6 to -1.2 V (SCE), the latter value depending upon the acidity of the solution.

Acidic conditions:

$$O_2 + 2H_3O^+ + 2e \longrightarrow H_2O_2 + 2H_2O$$

$$O_2 + 4H_3O^+ + 4e \longrightarrow 6H_2O$$

Neutral or alkaline conditions:

$$O_2 + 2H_2O + 2e \longrightarrow H_2O_2 + OH^-$$

$$O_2 + 2H_2O + 4e \longrightarrow 4OH^-$$

These reactions give rise to two kinds of problem. The most important of these is that the signal generated by the reactions (observed current) masks the required analyte reduction signals and renders the method useless in the range -0.5 V to at least -1.0 V (SCE). This is virtually the whole of the useful range of potential for reduction processes. The second type of problem arises from the direct interference of the reaction products, ie H_2O_2 and/or OH^-, with the analyte reaction. An example would be the effect of a localised rise in pH due to the generation of hydroxyl ions at the cathode.

It is thus very important to remove oxygen from cells in which cathodic processes are to be investigated.

∏   Describe how you would remove oxygen from the solution in an electrolysis cell.

This is done by passing oxygen free nitrogen through the solution for about 10 min. This leaves a residual amount of oxygen (a few ppm) and special steps must be taken to remove these remaining few ppm if necessary (3.2.4, Fig. 3.2h).

3.3.3.   Supporting Electrolytes

A supporting electrolyte is added to the solvent in all electrolysis-based analytical methods. These supporting electrolytes are strong electrolytes (strong in water) and are often referred to as indifferent or inert. These terms should be familiar to you from other areas of chemistry, eg the use of ion-selective electrodes. The terms indifferent or inert are used because the added ions must not participate directly in the required chemical reaction.

The supporting electrolyte has two roles to play.

(a) The concentration of the supporting electrolyte regulates the electrical resistance of the cell. Even with water, the most polar solvent used, the resistance across a liquid path of say 2 cm would be quite high. With aprotic solvents the electrical resistance is so high as to make electrolysis impractical unless a supporting electrolyte is added.

Calculate the resistance of a 2 cm length of a solution of 1.0 mol dm^{-3} $(Et)_4NBF_4$ in CH_3CN between two electrodes of areas 2 $\times$ 10^{-5} m^2. Conductivity of this solution is 5.55 S m^{-1}.

$$G = \text{conductance} = \kappa/J = \frac{5.55 \times 2 \times 10^{-5}}{0.02}$$

$$= 5.55 \times 10^{-3} \text{ S}$$

$$\therefore \quad R = 1/G = 180 \ \Omega$$

Compare this with question in 3.2.1.

(b) The supporting electrolyte controls the migration of ions between the working and the secondary electrode. You should know (1.3.1) that it is the mobility of an ion that determines the fraction of the total migration current carried by an ion and that the transport number of an ion is a measure of that fraction. In the bulk electrolyte the migration of ions under the potential gradient is the only net transfer process. Let us call the current so generated, I_m. Near the electrode, however, concentration gradients may occur particularly in unstirred solutions

(3.1.2). These gradients give rise to a diffusion current, I_d. It is of great importance in certain electro-analytical methods to understand the factors determining the relative magnitudes of I_m and I_d, eg in polarographic methods.

Consider a worked example.

An aqueous solution which contains $Cu(NO_3)_2$ of concentration 10^{-3} mol dm^{-3}, is electrolysed between a Pt anode and a Hg cathode at 25 °C. What are the relative magnitudes of I_m and I_d for the Cu^{2+} ion at the cathode? Mobility data at 25 °C: $u(Cu^{2+}) = 5.56 \times 10^{-8}$, $u(NO_3^-) = 7.4 \times 10^{-8}$ m^2 V^{-1} s^{-1}.

The total current at the electrode is $I = I_m + I_d$.

Let an arbitrary number of Faradays of electricity pass, say 10 Faradays (10 mol of electrons), ie I $\propto$ 10 units

$$t_i = \frac{z_i\, u_i\, c_i}{\Sigma z_i\, u_i\, c_i}$$

$t(Cu^{2+})$

$$= \frac{2 \times 5.56 \times 10^{-8} \times 10^{-3}}{(2 \times 5.56 \times 10^{-8} \times 10^{-3}) + (1 \times 7.4 \times 10^{-8} \times 2 \times 10^{-3})}$$

$$= \frac{1.112 \times 10^{-10}}{(1.112 + 1.48) \times 10^{-10}}$$

$$= 0.43$$

∴ A fraction 0.43 of the total cathodic current will be due to migration of ions to the electrode,

ie $I_m \propto 4.3$ units

∴ $I_d \propto 5.7$ units

∴ $I_m/I_d = 0.75$ for Cu^{2+} at the cathode

Π   An aqueous solution containing $Cu(NO_3)_2$ (10^{-3} mol dm^{-3})
          is electrolysed between a Pt anode and a Hg cathode at 25 °C.
          If KCl is added as a supporting electrolyte at a concentration
          0.1 mol dm^{-3} what are the relative magnitudes of I_m and I_d
          for the Cu^{2+} ion at the cathode? Additional mobility data at
          25 °C: $u(K^+)$ 7.6 × 10^{-8}; $u(Cl^-)$ 7.9 × 10^{-8} m^2 V^{-1} s^{-1}

This is a more complicated problem than the example in the text
but the principle is identical.

Answer is I_m/I_d = 0.007, ie > 99.3% diffusion.

Ion	$u_i\, c_i\, z_i$	
Cu^{2+}	5.56 × 10^{-8} × 10^{-3} × 2	= 1.112 × 10^{-10}
NO_3^-	7.4 × 10^{-8} × 2 × 10^{-3}	= 1.48 × 10^{-10}
K^+	7.6 × 10^{-8} × 0.1	= 7.6 × 10^{-9}
Cl^-	7.9 × 10^{-8} × 0.1	= 7.9 × 10^{-9}
	$\Sigma\, u_i\, c_i\, z_i$	= 1.576 × 10^{-8}

$$t(Cu^{2+}) = \frac{u_i\, c_i\, z_i}{\Sigma\, u_i\, c_i\, z_i} = \frac{1.112 \times 10^{-10}}{1.576 \times 10^{-8}}$$

$$= 0.007$$

∴   A fraction 0.007 of the total cathodic current will be due to
         migration of ions to the electrode.

ie I_m ∝ 0.07 units

∴ I_d ∝ 9.93 units.

∴ I_m/I_d = 0.007

A similar argument may be used for the processes occurring at the
other electrode in the cell.

The cathodic process here is clearly diffusion controlled.

We see that the addition of a supporting electrolyte at a concentration greatly in excess of that of the analyte causes the ionic migration of the analyte ion to be reduced to a negligible value. Usually the supporting electrolyte is added in at least ten-fold excess and preferably 100-fold excess with the limit often set by the solubility. Under these conditions the predominant mechanism for transport of the analyte ion to the electrode in an unstirred solution is diffusion. This is a very important fact in many voltammetric methods, eg polarography (3.5.1a).

If diffusion control is not essential then the supporting electrolyte concentration need only be sufficient to reduce the electrical resistance of the cell.

Some common supporting electrolytes used in aqueous or near aqueous solutions are:

KCl, $LiClO_4$, HCl, $HClO_4$, KOH, $NaOH$ together with buffer solutions based on weak organic acids and phosphates.

The common supporting electrolytes used primarily in polar aprotic and aprotic solvents are:

Tetraalkylammonium salts, $R_4N^+X^-$ where,

eg $\quad R = CH_3, C_2H_5$, t-butyl (C_4H_9), and

$\quad X = ClO_4^-, BF_4^-, Cl^-, Br^-, I^-$.

It is important for you to realise that the cathodic voltage limit is often set by the potential at which the supporting electrolyte cation, eg K^+, is reduced.

A note of caution on the use of supporting electrolytes when you are working at trace or lower levels of analyte concentration. When Analar grade supporting electrolytes are used at a concentration level of 0.1 mol dm^{-3} the concentration of impurities added inadvertently, eg heavy metals, is significant at the ppb level. This necessitates the use of a very pure and expensive grade of supporting electrolyte or the introduction of a purification procedure for the solvent/supporting electrolyte system. This purification is carried out electrochemically using a form of coulometry (4.2).

∏   Analar grade KCl is usually quoted as having <5 ppm Pb as an impurity. What is the concentration of Pb (in ppb) introduced into solution when 0.1 mol dm^{-3} aqueous KCl is used as a supporting electrolyte? [$A_r(K) = 39.1$, $A_r (Cl) = 35.5$]

The answer is 3.7×10^4 ppb or 0.037 ppm.

Assume that the manufacturer's statement is optimistic and that lead is present as the soluble Pb(II) ion at the level of 5 ppm in the solid KCl.

A 0.1 mol dm^{-3} solution of KCl contains 7.46 g KCl in 1 dm^3.

If Pb(II) is present at 5 ppm in the KCl then:

$$\text{Mass Pb(II) in 1 dm}^3 = \frac{7.46 \times 5}{10^6} \text{ g}$$

$$\text{Mass Pb(II) in } 10^6 \text{ cm}^3 = \frac{7.46 \times 5 \times 10^3}{10^6}$$

$$= 0.0373 \text{ g}$$

∴   Concenrtration of Pb(II) is 0.037 ppm or 3.7×10^4 ppb

3.3.4. The Voltage Window

You should be familiar with the concept of a spectroscopic window as used in the context of choosing a solvent or mulling agent for infrared spectroscopy. The aim is to find a solvent or mulling agent that is transparent to infrared radiation over as wide a frequency range as is possible.

We have here an analogous situation. The best choice of electrodes/solvent/supporting electrolyte is the one which shows no electrical activity, ie no appreciable change of current, over the widest range of WE potential. Into this voltage window may be fitted the changes of current due to a variety of analyte reactions of interest.

Consider some examples (approximate values of potential):

(a) $Hg/H_2O/1$ mol dm^{-3} $HClO_4$ $\qquad$ $+0.5$ to -1.0 V (SCE)

(b) $Hg/CH_3CN/0.1$ mol $dm^{-3}(C_4H_9)_4NClO_4$ $+0.5$ to -2.8 V (SCE)

(c) $Pt/H_2O/1$ mol dm^{-3} $HClO_4$ $\qquad$ $+1.3$ to -0.3 V (SCE)

(d) $Pt/CH_3CN/0.1$ mol dm^{-3} $(C_4H_9)_4NClO_4$ $+1.6$ to -2.5 V (SCE)

∏ $\quad$ What are the origins of the voltage limits stated in the above four examples (a) to (d)?

You should have answered as follows.

(a) -1.0 V (SCE) cathodic limit set by H_3O^+ reduction on Hg,

$\quad$ $+0.5$ V (SCE) anodic limit set by oxidation of Hg.

(b) -2.8 V (SCE) set by the reduction of $(C_4H_9)_4N^+$ on Hg,

$\quad$ $+0.5$ V (SCE) same as in (a).

(c) -0.3 V (SCE) set by H_3O^+ reduction on Pt,

$\quad$ $+1.3$ V (SCE) set by decomposition of solvent, O_2 evolved.

(d) -2.5 V (SCE) set by reduction of $(C_4H_9)_4N^+$ on Pt,

$\quad$ $+1.6$ V (SCE) set by formation of oxide film on Pt.

These examples cover most of the situations that decide voltage limits.

Remember that one rarely uses the same electrode material for both WE and SE.

Let us consider some more realistic systems:

(d) $H_2O/$ 1 mol dm^{-3} $HClO_4/Pt$(anode, SE)/Hg (cathode, WE);

(e) $H_2O/$ 0.1 mol dm^{-3} KCl/Pt(anode, SE)/Hg (cathode, WE);

(f) $CH_3CN/0.1$ mol $dm^{-3}(C_4H_9)_4NClO_4/Pt$ (anode SE)/C (cathode WE).

∏   What are the approximate voltage limits for the above examples (d), (e) and (f).

Example (d). Rearrange the data given for examples (a) and (c) to give: +1.3 to −1.0 V (SCE).

Example (e). Answer is +1.3 to −1.5 V (SCE). The anodic limit is the same as in (d) and (c). The cathodic limit is set by the reduction of K^+ ions.

Example (f). Answer is +1.6 to −2.6 V (SCE). The anodic limit is set as in (d). The cathodic limit you would have to estimate as slightly better than on a Pt electrode (d) and worse than on a Hg electrode (b).

3.3.5. Concluding Remarks

Given access to essential data you should be able now to design a suitable electrode/solvent/supporting electrolyte system for a particular purpose. Although you are now in a position to design a system to give a maximum voltage window, remember to keep an eye on convenience and cost, and do not over design. It is only necessary to provide a voltage window sufficient in size to accommodate the required analyte reaction/reactions. Some systems are akin to a comfortable old shoe. For example in the polarographic analysis of heavy metal cations the system, $H_2O/0.1$ mol dm^{-3} KCl/Pt(anode SE)/Hg(cathode, WE) is such a case with voltage limits +1.3 to −1.5 V (SCE). Most analysts would not change from this system for this type of analysis since wider voltage limits would have no advantages.

Be very careful with the above values of voltage limits. As stated they are the maximum positive value of potential that can be achieved at the anode to the maximum negative value that can be achieved at the cathode. In practice you are only interested in one electrode at any one time. For example, in polarography only the cathode (WE) potential is of interest and if this is Hg the useful range is about −0.1 V (SCE) to −1.5 V (SCE), hence the voltage window here is −0.1 V (SCE) to −1.5 V (SCE).

SAQ 3.3a　　What are the five essential features of a good electrode/solvent/supporting electrolyte system?

SAQ 3.3b　　What are the two roles of a supporting electrolyte? Calculate the relative magnitudes of the migration current and diffusion current for the zinc cation in an aqueous solution at 25 °C containing 5×10^{-4} mol dm^{-3} Zn(NO$_3$)$_2$ and 0.1 mol dm^{-3} KCl. Given mobility data (Fig. 1.3a):

Ion	$10^8 \, u/\text{m}^2 \, \text{V}^{-1} \, \text{s}^{-1}$
NO$_3^-$	7.4
Cl$^-$	7.9
K$^+$	7.6
Zn^{2+}	5.5

SAQ 3.3b

SAQ 3.3c   The standard reversible electrode potential for the reduction of hydronium ions in aqueous solution at 25 °C is −0.241 V (SCE) and yet hydrogen is evolved on the following surfaces at the potentials stated for a solution of 0.1 mol dm^{-3} aqueous HCl: Pt −0.321 V (SCE), Hg −1.363 V (SCE).

Explain this and state how this relates to the choice of electrode material for a cathode. [2.303 $RT/F = 0.0592$ V at 25 °C]

SAQ 3.3d	The following information is available for gold as a possible electrode material. The reduction of hydronium ions in 1.0 mol dm^{-3} aqueous $HClO_4$ occurs at about -0.5 V (SCE). Gold is as good as Pt on the anodic side provided complexing anions, eg CN^-, Cl^-, are absent. If these anions are present gold should not be used at potentials $> +0.5$ V (SCE). Summarise your conclusions about Hg, C, Pt and Au as materials to be used in the construction of working electrodes.

SUMMARY AND OBJECTIVES

Summary

The factors to be taken into account in choosing the electrodes, the solvent, and the supporting electrolyte for a particular working electrode reaction are discussed. You are now in a position to choose such a system, design a suitable cell, and describe a three-electrode circuit under either potentiostatic or galvanostatic control.

Objectives

You should now be able to:

- list the essential features of a good electrode/solvent/supporting electrolyte system and the ideal properties of a good reference electrode;

- describe the properties of mercury, platinum and carbon as electrode materials;

- explain the importance of the hydronium ion reduction and oxygen reduction reactions in electro-analytical chemistry;

- define anodic and cathodic voltage limits and voltage window;

- explain the roles of the supporting electrolyte in electrolysis-based electro-analytical methods and give examples of supporting electrolytes used in protic and aprotic solvents;

- carry out calculations to illustrate the effect of excess supporting electrolyte on the mechanism of ion transport and to determine the electrical resistance of typical solutions;

- calculate the concentration level of impurities introduced into solution by the addition of a supporting electrolyte;

- select suitable combinations of electrodes/solvent/supporting electrolyte to provide an adequate voltage window for a particular analysis.

3.4. INTRODUCTION TO VOLTAMMETRY AT FINITE CURRENT

You will see (4.1) that the naming of the various electro-analytical techniques has been largely a random matter. Only relatively recently have attempts been made to classify techniques and to show their inter-relationships.

The category of techniques now known as voltammetry at finite current includes the following methods.

Voltammetry at controlled potential

(*a*) dc polarography
(*b*) ac polarography
(*c*) solid electrode voltammetry
(*d*) amperometric titrations

Voltammetry at finite current (voltammetry)

(*e*) linear sweep voltammetry
(*f*) cyclic voltammetry
(*g*) stripping voltammetry
(*h*) advanced polarography
　(*i*)　sampled dc
　(*ii*)　normal pulse
　(*iii*) differential pulse

Voltammetry at controlled current

　chronopotentiometry

Note that having stressed 'at finite current' in order to distinguish from voltammetry at zero current, the finite current methods are commonly termed just voltammetry.

∏　Why was it necessary to define this category of techniques as voltammetry at finite current, ie why stress the 'finite current'?

We have just said that it was in order to distinguish these methods from voltammetry at zero current. These latter are the methods introduced in Section 2.8, ie potentiometry. You should be certain before proceeding that you clearly understand the experimental conditions used in potentiometry and those in electrolysis methods (3.1, 3.2, 3.3).

Voltammetry is usually taken to be the study of current/voltage relationships. More correctly it is the study of current/working electrode (WE) potential relationships.

3.4.1. Current/WE Potential Relationships

We have seen in Section 3.1 (Fig. 3.1a) that in an electrolysis cell the anode receives electrons from the solution (oxidation) and the cathode receives electrons from the external voltage source (reduction).

When an electrode is in equilibrium with a solution, with no external applied voltage, the electrode assumes a potential (E_e), the reversible electrode potential.

∏   For a reaction ox + ne $\rightleftharpoons$ red, what is the relationship between E_e and the concentrations of oxidised (ox) and reduced (red) states in the solution?

This is asking you to recall the Nernst equation from Section 2.4.

$$E_e = E^\ominus + (RT/nF) \ln [c(\text{red})/c(\text{ox})]$$

We have written this reversible electrode potential as E_e to distinguish it from a general electrode potential, E. Values of $E^\ominus$, the standard electrode potential, are always quoted on the SHE scale.

∏   For an aqueous cadmium sulphate solution (0.1 mol dm^{-3}), $E^\ominus(\text{Cd}^{2+},\text{Cd}) = -0.402$ V (SHE) at 25 °C. What is the value of $E_e(\text{Cd}^{2+},\text{Cd})$ on the SCE scale?

Answer is −0.673 V (SCE). If you did not obtain this answer study the following explanation.

$$E_e \text{ (SHE)} = E^\ominus - (0.06/2) \lg (1/0.10)$$
$$= -0.402 - 0.03 = -0.432 \text{ V (SHE)}$$
$$E_e \text{ (SCE)} = E_e \text{ (SHE)} - 0.241 \text{ V}$$
$$\therefore \quad E_e = -0.673 \text{ V (SCE)}$$

Revise Section 2.4 and subsection 3.2.2 if you are still having difficulties.

The Faradaic current is a measure of the rate at which an electrochemical reaction occurs and this rate is determined by two factors:

(a) the rate of the overall electron transfer process at the electrode surface, and

(b) the rate of movement of the electroactive species through the solution to the electrode, the rate of mass transport.

Considering each of these factors in turn.

(a) The electron transfer process

The current at the working electrode (WE), I, is given by

$$I = I_a + I_c,$$

where the relative magnitudes of I_a (anodic current) and I_c (cathodic current) reflect the extent to which oxidation and reduction are occurring at a particular electrode potential (E).

$\therefore$ When $E = E_e$, $I_a = I_c$, and $I = 0$, ie zero current conditions.

By convention the anodic current, I_a, is negative and the cathodic current, I_c, is positive.

When an external voltage is applied (electrolysis conditions) the electrode potential changes from E_e to E and a finite current develops.

$\therefore$ When $E_{app} > 0$, $I_a \neq I_c$, $I > 0$, $I = f(E)$.

In Fig. 3.4a (*i*) we see the typical $I = f(E)$ curves for a system where the electron transfer process is very fast. We see that as E moves away from E_e the current developed is either pure anodic or pure cathodic. These fast electron transfer processes occur when the activation energy barrier to the reaction is low, a situation common in metal/metal cation reactions.

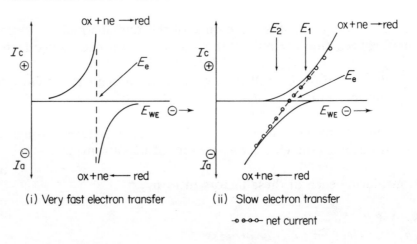

(i) Very fast electron transfer   (ii) Slow electron transfer

—o o o o— net current

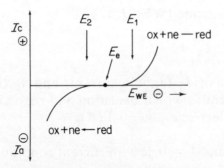

(iii) Very slow electron transfer

Fig. 3.4a. *Plot of I against E_{WE} for a pure electron transfer mechanism*

In Fig. 3.4a (*ii*) we see the typical $I = f(E)$ curves for a system where the electron transfer process is slow due to a high activation energy barrier for the reaction.

At $E = E_1$ we see that $| I_c | > | I_a$

$\therefore$ I is positive

At $E = E_2$ we see that $| I_c | < | I_a$

$\therefore$ I is negative.

In general for a slow electron transfer process there is a wide range of potential where the current is of mixed anodic/cathodic origin.

Π Explain what you understand by activation overpotential.

The essential points in answering this question are listed:

- if an electron transfer process is very fast then the potential at which the process occurs will be E_e, the reversible potential;

- if the activation energy barrier to electron transfer is high, additional potential is needed to achieve a finite rate and hence current, (Fig. 3.1b);

- overpotential, η, is a measure of the additional potential required.

$$\eta = E - E_e$$

If the activation energy barrier is very high then a situation arises as depicted in Fig. 3.4a(*iii*). The $I = f(E)$ curves are completely separated and it is not until E exceeds E_2 in the anodic direction or E exceeds E_1 in the cathodic direction that any reaction occurs and hence any current is observed.

The value $(E_1 - E_e)$ is the activation overpotential for the cathodic process and is negative.

The value $(E_2 - E_e)$ is the activation overpotential for the anodic process and is positive.

Reactions which fall into the categories illustrated in Fig. 3.4a (*i*) and (*ii*) are said to be reversible. Reactions which fall into the category illustrated in Fig. 3.4a (*iii*) are said to be irreversible.

(*b*) *Mass transfer processes*

As the electron transfer process becomes faster as a consequence of a more favourable electrode potential, a situation will eventually arise where the electroactive material is unable to reach the electrode at a sufficiently fast rate. We then find that the current reaches a limiting value dependent upon the rate of mass transport.

There are 3 mass transport mechanisms capable of transferring electroactive material to and from the electrode:

(*i*)   migration under the potential gradient;

(*ii*) diffusion under the concentration gradient;

(*iii*) convection due to stirring and/or thermal agitation.

During electrolysis and as a consequence of the electrode reaction, the concentration of electroactive species near to the electrode (the double layer) will change. A concentration gradient is created across the double layer and the thickness of the layer is dependent upon the stirring rate. The thickness is at a minimum for high stirring rates.

When a solution is quiescent (no stirring) we may assume that only factors (*i*) and (*ii*) are involved.

However, thermal agitation is present even in quiescent solutions and it is best to assume that a solution stays truly quiescent for no more than 5 minutes. Diffusion across a concentration gradient is usually the slowest mass transport process and is measured in terms of the diffusion coefficient, $D/m^2\ s^{-1}$. If diffusion becomes the dominant process, ie rate limiting, then the shape of the $I = f(E)$ curves change.

Π   How would you ensure that diffusion becomes the rate-limiting and hence the current-limiting process in an electrochemical reaction?

This is done by adding an excess of a supporting electrolyte to the solution. Revise Section 3.3.3 if you did not answer in this way or if you do not understand the principle.

Refer now to Section 3.1.2 and Fig. 3.1d and note the appearance of the curves relating current to overpotential (both concentration and activation types).

The characteristic feature of concentration overpotential is the limiting current, I_{lim}:

$$I_{\mathrm{lim}} = \eta FDcA/\delta \qquad (3.4a)$$

Where $c/\mathrm{mol\ m}^{-3}$ is the analyte concentration, A/m^2 the area of the electrode, and δ/m the thickness of the double layer.

Having considered the electron transfer and mass transport process separately let us now consider them together.

3.4.2.   Electron Transfer and Mass Transport

The curves that we have seen in Fig. 3.4a are idealised in the sense that mass transport was assumed to be infinitely fast, ie pure electron transfer control.

Consider a cathodic process:

$$\mathrm{ox} + \mathrm{ne} \longrightarrow \mathrm{red}$$

Fig. 3.4b shows the I_c/E curves when the behaviour is determined by both the electron transfer and the mass transport factors.

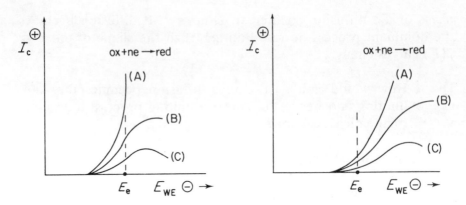

Fig. 3.4b. *Plot of I against* E_{WE} *for a reversible cathodic electrode process*

We see that the limiting current is the same in both cases but the slower the electron transfer process the more negative the potential that is required.

Does this make sense to you? As the electrode potential becomes more negative, even a naturally slow electron transfer process becomes faster until the mass transport limit is reached.

3.4.3. Analytical Implications

We now understand the factors affecting the shapes of *I/E* curves. How does this relate to analysis? It remains to devise suitable electrode/solvent/supporting electrolyte systems and operating conditions, ie analytical methods. The key to the successful method will be that the current developed at a particular point on the *I/E* curve will be reliably dependent upon the concentration of the analyte in the bulk electrolyte. Best of all will be a method where the current at some point is linearly dependent upon the analyte concentration.

SAQ 3.4a   List five techniques that belong in the category, voltammetry at finite current. What features do these techniques have in common?

SAQ 3.4a

SAQ 3.4b   State the main contribution to the Faradaic current in an electrolysis cell. How would you arrange for diffusion control of the cell current and how does this type of control manifest itself?

SAQ 3.4c	Draw the current/potential curves that you would expect for a reversible and irreversible reaction and explain the origin of the shapes of the curves.

SUMMARY AND OBJECTIVES

Summary

Voltammetric methods of analysis are classified into those occurring under controlled potential conditions and those occurring under controlled current conditions; eleven techniques are named.

The effects of the rate of the mass transport of analyte to the electrode and the rate of electron transfer at the electrode are discussed in terms of the resulting I/E curves and the relevance of this to analysis are established.

Objectives

You should now be able to:

• list the main techniques in the category voltammetry at finite current and place them into two sets; those at controlled potential and those at controlled current;

• state the convention for the polarity of anodic and cathodic current;

• list the types of mass transport that are relevant to electrochemical reactions and explain the origins of a limiting current;

• illustrate by sketches the effect of the rates of electron transfer and mass transport on the shape of I/E curves;

• explain the meaning of reversible and irreversible in the context of electrochemical reactions.

3.5. PRINCIPLES OF VOLTAMMETRIC METHODS

This section will not provide you with sufficient knowledge to be able to carry out any of the methods correctly. You will however be able to explain the essential principles of each technique, distinguish between them and to relate them to the analytical situation.

You may assume that all of these methods use three-electrode circuitry with potentiostatic or galvanostatic control. [This is a good time to revise Sections 3.2 and 3.3.]

3.5.1. Voltammetry at Controlled Potential

(a) dc Polarography

This method uses a dropping mercury electrode (DME) as a working electrode (WE). The DME consists of mercury drops generated at the lower end of a vertical glass capillary with a drop lifetime of

0.5 – 2 s. The secondary electrode (SE) is a platinum wire and the reference electrode (RE) is usually a saturated calomel electrode (SCE).

The electrode reaction is arranged to be under diffusion control by adding an excess supporting electrolyte and by using a quiescent solution. Oxygen is removed by purging with nitrogen. The potential of the WE is varied linearly and slowly at 2 to 10 mV s^{-1}.

For a cathodic process the *I/E* trace appears as in Fig. 3.5a. We see a polarographic wave, termed a polarogram.

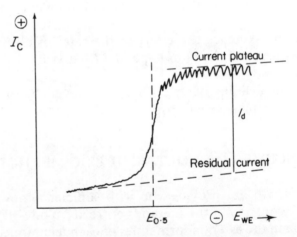

Fig. 3.5a.   *Typical polarogram, polarographic wave*

∏   What are the main features of a polarogram?

$E_{0.5}$ is the half-wave potential. The value of this quantity is characteristic of the electroactive species in a given solution (Fig. 3.5a). For a reversible reaction $E_{0.5} \simeq E^{\ominus}$, the standard reversible electrode potential. This quantity ($E_{0.5}$), is the basis of qualitative analysis by polarography.

I_d the diffusion current, is proportional to *c*, the concentration of analyte in the bulk solution. This quantity (I_d) is the basis of quantitative analysis in polarography.

Current plateau. At potentials on the current plateau a maximum (limiting) current has been reached.

Residual current. This is a non-Faradaic current which increases slowly as the potential increases in a negative sense. It is partly due to electroactive impurities but is mainly due to a capacitance effect arising at the mercury/solution interface.

Fig. 3.5b shows two waves and illustrates the principles of multi-component analysis.

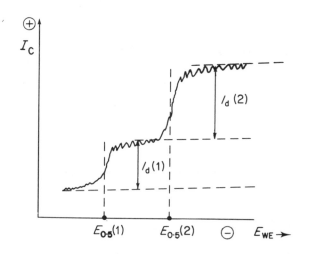

Fig. 3.5b. *Polarogram for a two-component system*

Π   Explain how a limiting current arises under the conditions described for dc polarography.

You should recall that the Faradaic current is made up of 2 factors, the electron transfer process and the mass transport process (3.4.1). As the rate of electron transfer increases, (in polarography this occurs as the potential becomes more negative), eventually the mass transfer process is unable to cope. Here the mass transport process is diffusion and this reaches a maximum rate which is dependent upon solution conditions.

Polarography is used almost exclusively for the analysis of reducible species and the chief category of analyte has been metallic cations. The detection limit is about 10^{-4} mol dm^{-3}.

$\Delta E_{0.5}$, ie the difference in the half wave potentials for neighbouring waves must be about 150 mV for their adequate resolution.

Π Why is polarography used almost exclusively for analyses involving cathodic reactions?

You have learnt (3.3) that mercury is a good electrode material for cathodic (reduction) reactions but very poor for anodic (oxidation) reactions.

The key words here are voltage limits and voltage windows for electrode/solvent/supporting electrolyte systems. You should be able to discuss the relative merits of Hg, Pt and C as electrode materials.

(b) ac Polarography

The experimental conditions are as for dc polarography except that a small ac voltage is superimposed on the slow linear dc voltage ramp. Special equipment is required to isolate the alternating component of the cell current. It is this ac component of the current which is displayed against E(WE).

Fig. 3.5c shows a typical ac polarogram for a two component mixture.

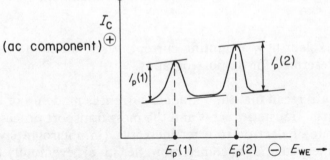

Fig. 3.5c.   *Typical ac polarogram*

I_p is the current at the peak maximum; $I_p \propto c$

E_p is the potential at the peak maximum; E_p has the same significance as $E_{0.5}$

In favourable cases the detection limit is about 10^{-6} mol dm/$^{-3}$. One important fact is that the solutions need not be purged of oxygen. This is because the ac method is insensitive to irreversible electrode processes such as the oxygen reduction reaction.

ΔE_p for the resolution of neighbouring peaks needs to be about 50 mV.

(c) Solid Electrode Voltammetry

For this technique the WE is usually platinum, gold or carbon. The platinum or gold is sealed into a glass rod and carbon is used in rod form. The WE is usually rotated at a uniform rate, typically 600 rpm. The solution is purged of oxygen and a supporting electrolyte is present. Clearly the solution is stirred by the moving electrode. Fig. 3.5d shows a typical voltammogram.

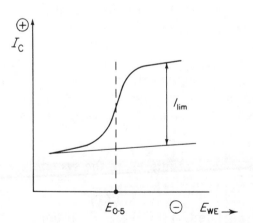

Fig. 3.5d. *Typical solid electrode voltammogram*

I_{lim} is the limiting current and provided that the stirring rate is constant; $I_{lim} \propto c$.

At about 600 rpm the sensitivity is about five times better than that in dc polarography. Oxidation reactions may be studied as readily as reduction reactions (compare polarography). One disadvantage of the method is the susceptibility of solid electrodes to contamination due to adsorption effects. This tendency also makes multicomponent analysis less reliable, ie the first analyte to be deposited affects the deposition of other analytes.

∏   In solid electrode voltammetry what are the factors determining the value of I_{lim}?

Since a supporting electrolyte is added and the solution is rapidly stirred, diffusion and forced convection are the mass transport processes. The more rapid the stirring, the smaller the diffusion layer and consequently the higher the value of I_{lim} (3.4a).

So the factors determining I_{lim} are the stirring rate, the diffusion coefficient of the analyte and the concentration of the analyte.

(d) Amperometric Titrations

The solution and electrode conditions used are the same as those used for dc polarography or for solid electrode voltammetry. In addition provision is made to add a reactant, the titrant, to the cell in a controlled manner, eg, from a burette. The potential of the WE remains constant throughout the analysis and the value of E_{WE} is chosen so that at least one species in the system is electroactive. Usually the choice of potential is made on the basis of known polarographic data.

We have three species of interest in the system:

$$\text{analyte (A)} \xrightarrow{\text{titrant (T)}} \text{product (P)}$$

The current is displayed as a function of the volume of titrant added.

The appearance of the result will depend upon the electroactivities of the three species at the chosen potential, measured by their respective I_d values (compare conductimetric titrations and the molar ionic conductance, $\lambda \propto$). The end-point of the titration gives the analytical result. Fig. 3.5e (*i*) and (*ii*) show two examples.

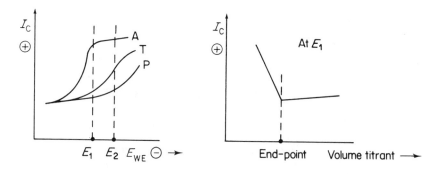

(i) Analyte very electroactive, titrant and product feebly electroactive

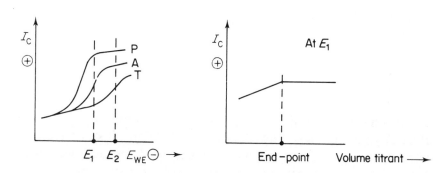

(ii) Product very electroactive, analyte less electroactive, titrant feebly electroactive

Fig. 3.5e.   *Typicalamperometric titrations*

The sensitivity is comparable with that of dc polarography. Using solid electrodes, potentials may be used where oxygen is not reduced thus removing the necessity for purging of the solution.

∏   Sketch the appearance of the titration plot if the fixed po-
     tentials (Fig. 3.5e, (*i*), and (*ii*)) were moved from E_1 to E_2.

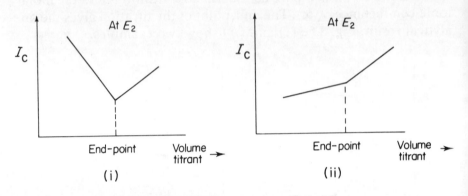

Fig. 3.5f.   *Amperometric titration result at potential E_2*

In (*i*) at potential E_2 both analyte and titrant are electroactive.

In (*ii*) at potential E_2 all three; titrant, analyte and product are
electroactive.

(*e*) Linear Sweep Voltammetry

A stationary working electrode is used. This may be Pt, Au, C or
the hanging mercury drop electrode (HMDE). This latter electrode
consists of a single drop of mercury extruded on the lower end of a
vertical glass capillary. The remaining solution conditions are as for
dc polarography. A rapid linear potential scan is used, in the range
20 to 400 mV s^{-1} and at the fastest rate a cathode ray oscilloscope is
required to capture the result. The electrochemical reaction is diffusion controlled. Fig. 3.5g shows the appearance of a linear sweep
voltammogram.

E_p is the peak potential given by

$$E_p = E_{0.5} - 0.028/n \text{ V at 25 °C.}$$

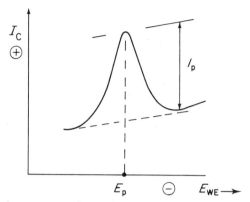

Fig. 3.5g. *Typical linear sweep voltammogram*

I_p, the current at the peak potential is proportional to $cv^{1/2}$ for a reversible process, where v/V s^{-1} is the scan rate.

At constant scan rate, $I_p \propto c$. Even for an irreversible process, $I_p \propto c$ but the proportionality constant is lowered.

The detection limit is about 10^{-5} mol dm^{-3} and the method is very fast. For good resolution of neighbouring peaks, ΔE_p must be about 100 mV. Usually only the first peak in a multicomponent system is reproducible due to contamination of the electrode by the first species deposited.

∏   Attempt to explain the peak appearance in linear sweep voltammetry compared with the wave appearance in dc polarography.

By now we have dealt with the idea of electron transfer and mass transfer processes determining the value of the observed current. The additional factor to take into account here is the rapidity of the potential scan, >20 mV s^{-1}. The current rises rapidly at first but the analyte concentration at the electrode surface is also rapidly depleted and the current begins to fall to reach a diffusion limited value. The theory of the shape of I/E curves giving rise to the wave shape applies only to quasi-equilibrium conditions. These conditions begin to break down when scan rates >10 mV s^{-1} are used. The electrode/bulk electrolyte concentration gradients do not have time to keep up with the changing potential.

(f) Cyclic Voltammetry

Cyclic voltammetry is an extension of linear sweep voltammetry in which the potential of the working electrode is swept first in one direction and is then reversed (Fig. 3.2d). The sweep rates are fast, 20–400 mV s^{-1}. Fig. 3.5h shows a typical cyclic voltammogram for one electroactive species in solution.

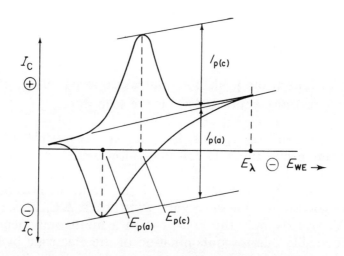

Fig. 3.5h.   *Typical cyclic voltammogram*

For a reversible process, $I_{p(a)} = I_{p(c)}$, and both of these quantities are proportional to c, the analyte concentration.

E_λ is the switching potential, the potential at which the scan is reversed.

$$\Delta E_p, \text{ie } E_{p(c)} - E_{p(a)} = 59/n \text{ mV at } 25 \text{ °C.}$$

For analysis there is no advantage over linear sweep voltammetry but it is an extremely powerful tool for mechanistic studies and for the identification of short-lived intermediates.

(g) Stripping Voltammetry

We shall discuss this method in terms of anodic stripping voltammetry (ASV). The experimental conditions are as for linear sweep voltammetry except that almost always a mercury working electrode is used. This takes the form of a hanging mercury drop electrode (HMDE) or a mercury film electrode (MFE). The latter electrode is discussed in 3.3.1.

The analyte is first reduced at a fixed potential at the WE with stirring and deposits on the mercury. Since the volume of the mercury is small compared with the volume of the solution, this step amounts to a concentration of the analyte; this is the deposition stage. After a short rest period with no stirring, the equilibration stage, the electrode is now subjected to a fast anodic potential scan $(20–100 \text{ mV s}^{-1})$ during which time the analyte is oxidised back into the solution; this is the stripping stage.

The fixed potential (E') in the deposition stage is chosen to be well on to the polarographic current plateau of the analyte, Fig. 3.5i (i). The current developed in the stripping stage is related to the potential as in linear sweep voltammetry, Fig. 3.5i (ii).

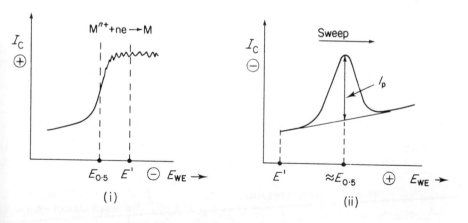

Fig. 3.5i. *Stripping voltammetry (i) polarogram showing choice of potential (E') for deposition stage, (ii) anodic sweep in stripping stage.*

$I_p \propto c'$ (analyte concentration on the electrode)

$\propto c$ (analyte concentration in the bulk solution)

This method is used extensively for the analysis of very dilute solutions of metallic cations. The detection limit is about 10^{-8} mol dm^{-3} (0.001 ppm). Multicomponent analysis can be carried out quantitatively, Fig. 3.5j. At these low analyte concentration levels impurities in the solvent/supporting electrolyte system become a problem. Refer back to subsection 3.3.3 for a discussion of this problem.

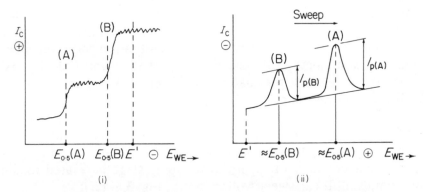

Fig. 3.5j. *Multicomponent stripping voltammetry*
(i) polarogram showing choice of potential (E') for deposition stage, (ii) anodic sweep in stripping stage.

∏   By analogy with the description of anodic stripping voltammetry, describe how you would carry out an experiment that could be termed cathodic stripping voltammetry.

The approach is identical to that for ASV. First there is a deposition stage. It will be necessary to choose a E_{WE} at which the anodic (oxidation) deposition step may occur.

You know that this may preclude the use of mercury as the electrode material; it will if $E_{WE} > 0.2$ V (SCE). After the deposition stage and a short equilibration period a fast cathodic sweep stripping stage is carried out. The result will appear as in Fig. 3.5i (*ii*) except that the potential sweep will be cathodic and the current measured will be cathodic (positive).

(*h*) *Advanced Polarographic Methods*

We have seen (3.5.1a) that in dc polarography there are two contributions to the total current; the Faradaic current and the residual current, with the latter being due mainly to a capacitance effect at the mercury/solution interface. It is the magnitude of this capacitive current which sets the detection limit for the dc method. At about 10^{-4} mol dm^{-3} of analyte the capacitive current becomes comparable with the Faradaic current and the wave shape disappears.

The advanced methods all, in one way or another, discriminate between the Faradaic and capacitive currents and in favour of the Faradaic contribution. In Fig. 3.5k we see that the time dependance of the Faradaic current I_f and of the capacitive current I_{cp} during the lifetime of a mercury drop. We see that $(I_f - I_{cp})$ is at a maximum in the last few seconds of the drop lifetime. Advantage is taken of this fact in all of the following methods.

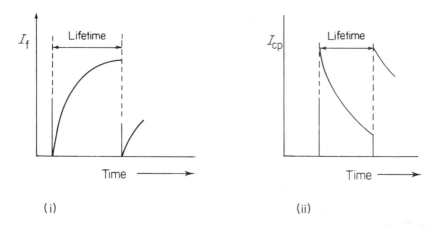

(i)   (ii)

Fig. 3.5k. *Faradic (i) and capacitive (ii) currents during the lifetime of a mercury drop*

The conditions for these methods are the same as for those described in dc polarography with two exceptions, firstly, specialised circuitry is required to process the measured current values and secondly the drop time must be very reproducible. The latter requirement necessitates the use of a mechanical tapper to dislodge the drop and the tapper is actuated by the electronic circuitry. This enables the drop time to be synchronised with other processes.

1. Sampled dc polarography (Tast polarography)

The current is sampled only during the last few seconds of the drop lifetime. This gives a curve similar in appearance to that for dc polarography but much smoother and with enhanced sensitivity, Fig. 3.51. The detection limit is about 10^{-6} mol dm^{-3}.

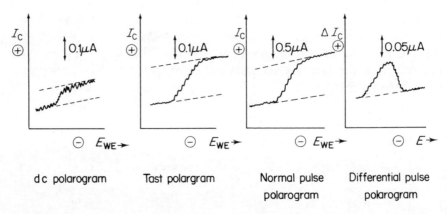

dc polarogram   Tast polargram   Normal pulse polarogram   Differential pulse polarogram

Fig. 3.51. *Types of polarograms, all for 10^{-5} mol dm^{-3} analyte*

2. Normal pulse polarography

Consider a cathodic (reduction) process. The WE potential is set just too positive to achieve any electrochemical reaction and is maintained at this value.

A potential pulse of a few mV is superimposed on this base potential near the end of the drop lifetime. The current is sampled at the end of the pulse, which coincides with the end of the life of the drop. The additional potential, in the form of the pulse, is increased each time in a negative direction taking the total potential into the electroactive range of the analyte. Only the sampled current is recorded. The pulse enhances I_f relative to I_{cp} and sampling at the end of the drop lifetime discriminates against I_{cp} as in Tast polarography. This gives a wave as before but with a characteristic step form, Fig. 3.5l. The detection limit is lowered to about 10^{-6} to 10^{-7} mol dm^{-3}.

3. Differential pulse polarography

The dc voltage ramp of 2 to 10 mV s^{-1}, used in dc polarography, is also used here. A small fixed potential pulse (in the range 10 to 100 mV) is applied near the end of the drop lifetime. The current is sampled twice, once just before the pulse is applied and again at the end of the pulse (just before the drop falls). The electronic circuitry enables the difference between these two currents to be calculated and plotted. The greater the pulse (in mV) the greater the sensitivity but at the same time the peak obtained broadens, lowering the resolution between neighbouring peaks. The form of the result is a peak not a wave, with the pulse steps clearly visible. Fig. 3.5l. The detection limit is about 10^{-8} mol dm^{-3}.

∏   If in the techniques 1 and 2 the current is sampled during the last 10 ms of the drop lifetime and that lifetime is 0.5 to 2 s, what may we say about the size of the drop when the current is sampled?

The answer required is that the drop is the same size every time sampling occurs. The electrode is to all intents and purposes, a fixed area electrode, and importantly new and clean for each drop. In technique 3 the pulse lasts for about 50 ms so the first current sampled is on an electrode of smaller size than that when the second current is sampled. However, we are still very near the end of the drop lifetime and the area will not change appreciably. More important the two areas are reproducible from drop to drop.

We have seen that the mercury drop is virtually a fixed area electrode in the above three methods. When differential pulse methods are applied to the stripping stage in stripping voltammetry we have differential pulse anodic stripping voltammetry (DPASV). The detection limit here can be as low as 10^{-11} mol dm^{-3}, ie, the ppb level. This makes the technique competitive with atomic absorption spectroscopy for the analysis of cations of the heavy metals.

3.5.2. Voltammetry at Controlled Current

Any method requiring controlled current conditions will require galvanostatic control (3.2.3). The only technique in this category that we need introduce is that of chronopotentiometry. In this method the current is maintained at a constant value and the working electrode potential is monitored as a function of time.

Although a hanging mercury drop electrode or a mercury film electrode could be used, the working electrode is usually Pt, Au or C. A supporting electrolyte is used and oxygen must be removed if reduction processes are to be studied. The solutions are quiescent and diffusion is the rate limiting process.

Fig. 3.5m shows a typical result when E_{WE} is plotted against time, where τ is the transition time and $E_{\tau/4} \simeq E_{0.5} \simeq E^{\ominus}$ is characteristic for the reaction occurring.

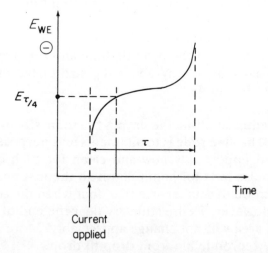

Fig. 3.5m.   *Chronopotentiometry, typical result*

The *Sand* equation applies and this shows that:

$\tau^{1/2} \propto cA/I \propto c$ (at constant current)

A is the area of the electrode.

Since the method relies on quiescent conditions, the current is usually adjusted so as to make the transition time of the order of a few seconds to one minute. The sensitivity is only as good as for dc polarography, about 10^{-4} mol dm^{-3} is the detection limit.

In principle the method could be used for multicomponent analysis provided $\Delta E_{\tau/4} > 150$ mV. However, as with all solid electrode techniques, the quantitative determination is good only for the first species deposited due to adsorption effects.

SAQ 3.5a Describe the principles of an electro-analytical method suitable for the determination of lead at the ppb level.

SAQ 3.5b What is the main factor that determines the limit of detection in dc polarography? How, in general, are the more advanced forms of polarography designed to overcome this factor?

SAQ 3.5b

SAQ 3.5c   Describe a suitable electro-analytical method
               that could be used to determine nitrobenzene
               in acetonitrile solution at a concentration level
               of about 10^{-7} mol dm^{-3}.

SAQ 3.5d

The voltammograms for Br_2,Br^- and Sn^{4+},Sn^{2+} in aqueous solution at 25 °C obtained on a platinum working electrode are shown in Fig 3.5n.

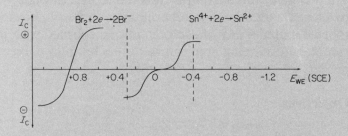

Fig. 3.5n. *Solid electrode voltammograms*

Draw the resulting amperometric titration result when a solution of tin(II) ions (Sn^{2+}) is titrated with bromine (Br_2) at a fixed working electrode potential of: (*i*) $E_{WE} = +0.3$ V (SCE), (*ii*) $E_{WE} = -0.4$ V (SCE)

SUMMARY AND OBJECTIVES

Summary

The whole of this section is devoted to describing the main features of eleven voltammetric techniques. You will, however, require a more detailed study of each technique coupled with practical experience before being able to use the techniques correctly.

Objectives

You should now be able to:

● describe the essential features of the main techniques of voltammetry at finite current;

● explain the following terms that arise in the description of the techniques; polarogram, polarographic wave, half-wave potential, diffusion current, current plateau, residual current, current at peak maximum, potential at peak maximum, switching potential, transition time.

4. Review of Methods of Electro-analytical Chemistry

Overview

In this part a classification of the methods of electro-analytical chemistry is presented. Potentiometric and voltammetric methods already introduced are referred to and their place in the classification identified.

The underlying principles of further techniques, principally coulometric techniques, are presented to complete the introduction to all branches of the classification.

Finally, the more important techniques in terms of present day usage are identified and attention is drawn to further study possibilities.

4.1. CLASSIFICATION AND RELATIONSHIPS

One of the most confusing aspects of electro-analytical chemistry is that there are so many different methods. Some of these methods differ only in very few respects but often the differences are important and require a good understanding of principles to be appreciated. The methods have not sprouted logically over the years from a single stem of electrochemistry. Their names have been given since the early nineteenth century with little or no semblance of a systematic approach.

It is only in recent years that there has been any attempt to bring some coherence into the nomenclature of the subject. This has been done, not by changing established names, but by placing the techniques together in logical sets, ie classification systems have been proposed. It is unfortunate that there is still no universally accepted classification but the one given below is widely used.

For almost all of the methods of electro-analysis, we are interested in the relationships between four variables: current, working electrode potential, analyte concentration and time. The only important exceptions to this are the techniques of electrogravimmetric analysis and those techniques based on solution conductance, principally conductimetric titrations. Electrogravimmetric analysis will be discussed briefly later (4.2) The methods based on solution conductance are assumed to have been covered in previous studies. If not refer to Fifield and Kealey (1983).

∏   Name the broad classifications of electro-analytical techniques that you have already encountered.

On the broadest level of analytical chemistry we have destructive and non-destructive methods. All analytical methods can be placed in one or other of these categories.

Electrolysis based electro-analytical methods may be classified as microelectrolysis or macroelectrolysis methods. Generally speaking all of these methods are non-destructive. The analyte is almost completely or completely removed during macroelectrolysis methods but is usually recoverable from the working electrode. In microelectrolysis the analyte solution remains essentially unchanged.

We may classify the methods as follows:

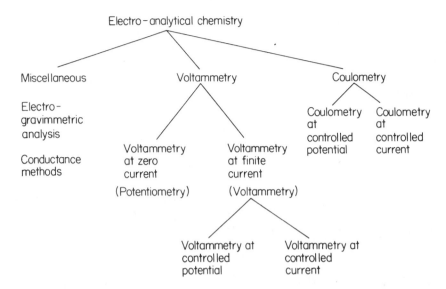

We have already introduced the most important methods in the categories voltammetry at finite current (3.5) and voltammetry at zero current, potentiometry, (2.8).

The only remaining group of methods is that based on Faraday's laws. These methods depend upon the relationship between concentration changes and the quantity (ie coulombs) of electricity used- hence coulometry (4.2.2).

If we inspect the classification and look in particular at the names of individual techniques (3.5), we see clearly some of the reasons for confusion. Perhaps the best example is chronopotentiometry (named in 1901) which is classified (correctly) as voltammetry at finite current with controlled current conditions, ie voltammetry at controlled current. It is not a member of the potentiometry family.

Π   What is it that all the techniques with the word polarography in their name have in common?

The word polarography should be reserved for those voltammetry techniques at finite current where the working electrode is a dropping mercury electrode. Similar methods, using working electrodes made of other materials, should be called voltammetry, eg solid electrode voltammetry.

In the literature of the subject you will find alternative classifications and in some of these the term amperometry appears. The only residual use of this term in our classification is in the technique amperometric titrations (3.5.1d) You have been introduced (2.3) to the recommendations of the Internation Union of Pure and Applied Chemistry (IUPAC); in that case nomenclature for reversible electrochemical cells. IUPAC have produced a classification for electroanalytical techniques –IUPAC, Classification and Nomenclature of Electroanalytical Techniques (Rules Approved 1975), Pergamon, 1976. There is a strong resemblance between these recommendations and our classification. The agreement however is not exact and the exceptions have been made where customary, contemporary usage differs still from the IUPAC rules.

SAQ 4.1a	Sketch out a classification of the methods of electro-analytical chemistry and correctly place the following techniques, cyclic voltammetry, chronopotentiometry, ion-selective electrode methods.

SUMMARY AND OBJECTIVES

Summary

An example of a classification of electro-analytical techniques is presented and the more important techniques fitted into this classification. The literature of the subject is still not presented in a uniform manner and reference has been made to IUPAC recommendations.

Objectives

You should now be able to:

* state the four variables whose inter-relationships form the basis of most electro-analytical methods;

* draw a diagram showing the main sub-divisions of the electro-analytical family of methods and give examples of techniques in the various sub-divisions.

4.2. PRINCIPLES OF THE MORE IMPORTANT METHODS

Before studying this section you are advised to revise previous sections on potentiometry (2.8) and voltammetry (3.5)

There remains only electrogravimetric analysis and coulometric methods.

4.2.1. Electrogravimetric Analysis

This method uses the three-electrode circuitry already described (3.2). The main application is to the analysis for metal cations hence the method is almost always used in the reduction mode, ie with a cathodic working electrode. Complete and exclusive removal of the analyte is required and the reduced form of the analyte is deposited (plated) on to the solid working electrode. In order to sustain the level of current required, the working electrode must have a large surface area and, typically a cylindrical platinum mesh electrode is used. A small platinum secondary electrode completes the circuit and this electrode is ideally isolated in its own compartment separated from the working electrode compartment by a porous glass frit. This isolation of the secondary electrode is not always a necessary design feature, it depends upon the likelihood of interference of products of the secondary electrode reaction with the required working electrode reaction. The whole solution is stirred during analysis. Controlled potential conditions are preferred and the potential of the working electrode is chosen to be on the current plateau of the polarogram of the analyte.

The quantitative measure is the weight of material deposited. For this reason the method is not a trace analysis technique, it is best used when the analyte is present at 5% w/v or greater. A Faradaic efficiency of 100% is assumed.

The method finds wide use as a separation method often using a mercury-pool cathode of large area. Impurities are removed by selective electrolysis at controlled potential. We have described the use of this method to purify supporting electrolyte solutions in (3.3.3).

∏   The $E_{0.5}$ values for:

Cd^{2+} + 2e $\longrightarrow$ Cd and Zn^{2+} + 2e $\longrightarrow$ Zn are respectively −0.64 V (SCE) and −1.10 V (SCE) in 0.1 mol dm^{-3} KCl as supporting electrolyte. Show by a sketch, how you would select the values of E_{WE} for the deposition of Cd followed by the deposition of Zn.

If the value of E_{WE} is set much less negative than $E_{0.5}$ then that reduction process will not occur. The polarogram for this 2-component system is shown in Fig. 4.2a.

If we select E_{WE} at about −0.8 V (SCE), 1, then only Cd will deposit. When all the Cd has deposited change E_{WE} to about −1.3 V (SCE), 2, and the Zn will deposit. If necessary the working electrode could be replaced between the two deposition stages.

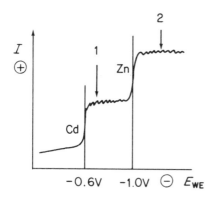

Fig. 4.2a. *Polarogram for Cd and Zn*

∏ What value of E_{WE} would you choose in order to remove all possible interfering reducible species from a 0.1 mol dm^{-3} aqueous KCl solution to be used as a supporting electrolyte solution for polarography?

We know that the K^+ ion is itself reduced at about -1.6 V (SCE). Choose, therefore, a potential close to this value but less negative, say about -1.4 V (SCE). This will certainly cause the heavy metal impurities present even in Analar grade KCl to be deposited.

4.2.2. Coulometric Methods

Two categories of method are possible, either coulometry at controlled potential or coulometry at controlled current. All the methods are based on Faraday's laws and all require three-electrode circuitry (3.2)

∏ What do you understand by the statement that an electrode reaction occurs with 100% Faradaic efficiency?

If the only electrode reaction occurring is the one required, ie the analyte reaction, then we have 100% Faradaic efficiency. When the overall composition of a solution is unknown it may be that under the experimental conditions more than one electrode reaction will occur. This will cause the weight deposited in electrogravimetric analysis to be too high. An error is also introduced into coulometric techniques.

(a) Coulometry at Controlled Potential

The conditions are very similar to those used for electrogravimetric analysis at controlled potential. The working electrode is usually a cylinder of platinum mesh, large in area by polarography standards but much smaller than in electrogravimetric analysis. It is necessary in this technique, to place the secondary electrode in a compartment separated from the rest of the cell by a porous glass frit. The solution is de-oxygenated if reduction processes are to be studied and the solutions are stirred. The potential of the working electrode is chosen on the current plateau of the analyte polarogram. Fig. 4.2b shows a typical cell design.

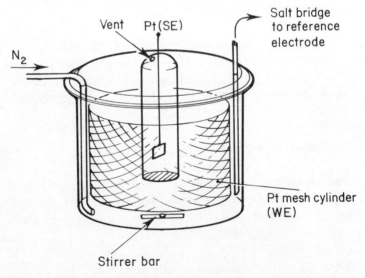

Fig. 4.2b.   *Coulometry cell*

The initial current can be in the range of a few μA to 100 mA and the current decreases with time. The current is recorded as a function of time and a typical decay curve is shown in Fig. 4.2c.

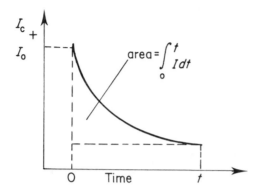

Fig. 4.2c. *Current decay in controlled potential coulometry*

$$I_t = I_0 \exp(-pt)$$

where p, the mass transfer factor, is given by mA/V,
A/m^2 is the area of the working electrode,
V/m^3 is the volume of the solution and,
m/ms^{-1} is the mass transfer coefficient.

Good cell design maximises p by using an electrode with a large area in a cell of small volume, with a large mass transfer factor achieved by efficient stirring of the solution. With p as large as possible analysis times are reduced, Fig. 4.2d.

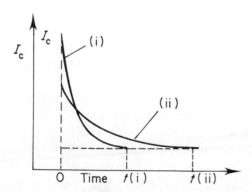

Fig. 4.2d. *Effect of cell design on controlled potential coulometry*
(i) p large; (ii) p small; areas under curves equal

Faraday's law gives the relationship:

$$\text{mol electrochemical reaction} = Q/nF$$

where Q coulombs is the amount of electricity passed.

$$Q = \int_0^t I\,\mathrm{d}t$$

Clearly Q is the area under the I/t curve and is proportional to the concentration of the analyte. For quantitative analysis it is necessary to measure the area under the I/t curve. The integration of the data curve is achieved as part of the instrumentation and a digital print-out is presented which is proportional to the area. The instrument is usually set to stop measurement of the area when $I_t < 0.1\% \; I_0$.

A multicomponent system can be analysed provided $\Delta E_{0.5} > 200$ mV, by changing the potential of the working electrode. Errors arise if the process has less than 100% Faradaic efficiency. In principle standard solutions are not required. The detection limit in favourable circumstances and with good quality electronics can be as low as 10^{-8} mol dm^{-3}.

(b) Coulometry at Controlled Current

The most often used method in this category is coulometric titrations. In this method a reagent (titrant) is generated quantitatively at the working electrode. A suitable reagent precursor is required. For example if bromine is required as a reagent, a soluble bromide is a suitable precursor. A supporting electrolyte is used and oxygen is removed if reduction processes are involved. Small platinum electrodes are usually used for both WE and SE and the secondary electrode must be placed in a compartment separated from the working electrode. A fixed current is applied for a known time in a stirred solution, the level of current being typically 10 μA to 100 mA and the time 10 to 100 s. A given current for a given time will generate a known amount of reagent, eg 10 μA for 10 s generates about 10 ng of bromine from an aqueous bromide solution. This reagent may be used for any normal titration purposes, ie the circuitry is simply replacing a burette. It is necessary to incorporate into the cell provision for the detection of the end-point of the titration. This could be a colour-change indicator, or a pH-measurement system for an acid-base titration, or the electrolysis cell could be designed to fit into the cell compartment of a uv-visible spectrophotometer.

This is an extremely useful titration method when the reagent generated (titrant) is unstable or in any way difficult to store. The amount of titrant generated can be very small but is still generated in a controlled manner. No primary standards are required.

SAQ 4.2a	Describe how you would remove heavy metal impurities from a 0.1 mol dm^{-3} aqueous solution of potassium chloride.

SAQ 4.2b An organic ketone may be determined quantitatively using bromine as a standard. Describe how you would determine such a ketone at the ppm level in aqueous solution using an electroanalytical method.

SAQ 4.2c The phenol content of a river water sample is required. A 20 cm^3 sample is made slightly acidic and an excess of potassium bromide is added. It required a steady current of 8.6 mA for 187 s to produce the bromine required to quantitatively react with phenol by the reaction:

$$PhOH + 3\,Br_2 \longrightarrow C_6H_2(OH)Br_3 + 3\,HBr.$$

What is the concentration of phenol in the water in ppm assuming that the river water has a density of 1.00 g cm^{-3}?

$[F = 96485 \text{ C mol}^{-1}; M_r(PhOH) = 94.1]$

SAQ 4.2c

SUMMARY AND OBJECTIVES

Summary

The more important techniques under the headings, potentiometry and voltammetry had been introduced earlier in this unit. This section deals with electrogravimetric analysis and the coulometric methods, and so completes the description of the members of the electroanalytical methods family except for conductance methods. It has been assumed that conductimetric titration techniques are known to you but a reference is given in case supplementary reading is required.

Objectives

You should now be able to:

- describe the techniques: electrogravimetric analysis, coulometry at controlled potential and coulometric titrations.

4.3.   PRESENT USAGE

In this unit we have introduced about 17 named techniques. It remains to comment on the extent to which these techniques are used. Which are the essential methods, the ones most likely to be encountered in industry?

The first choice is most likely ion-selective electrodes, the most recent of the potentiometric techniques and the technique where the most rapid developments are taking place. These developments are especially interesting in the fields of biochemistry, pharmaceutical and medicinal chemistry.

Next would be the family of polarographic methods and, in particular, the most recent of these using pulse techniques. Modern developments have been directed towards improved automation. The most active growth point in applications is to the pharmaceutical area. Equally important, and related to polarography, is the technique of stripping voltammetry. Using pulse techniques this method is now a serious competitor for applications to the analysis of metallic cations at the ppb level.

The remaining techniques all have their place for certain situations, certain problems, but are not used routinely in modern analytical laboratories. This introductory unit together with the bibliography will enable you to understand the principles of the methods and to find further information for yourself.

Self Assessment
Questions and Responses

SAQ 1.1a   Consider the first layer of solvent molecules around an ion. Which of the items below do you think are important in determining the number of solvent molecules in this layer?

(*i*)   ionic size;
(*ii*)   ionic charge;
(*iii*)   solvent structure;
(*iv*)   solvent molecular volume.

Response

Your answer should be (*i*), (*iii*), (*iv*).

The packing of the solvent molecules around an ion can be considered as mainly a geometric packing problem this being governed by the relative sizes of the ion and the solvent molecule (ie by (*i*) and (*iv*)). This must be modified however by the requirement that the solvent molecule must orientate itself toward the ion in such a way that ion-dipole or hydrogen bonding forces are maximised (ie (*iii*)). Ionic charge (*ii*) affects the size of the ion-solvent molecular force and not normally the number of solvent molecules around the ion.

****

SAQ 1.1b   Classify the following solvents as associated or otherwise: benzene; ethanol; trichloromethane.

Response

Benzene is non-associated.

Ethanol is associated; it has the possibility of forming hydrogen bonds.

Trichloromethane is non-associated.

****

SAQ 1.1c   Explain with reasons which of the following molecules you would expect to have a zero dipole moment: H_2; H_2O; Ar; NH_3; CCl_4; $CHCl_3$; CO; CO_2; C_6H_6; $C_6H_5NO_2$.

Response

Dipole moments arise because of the difference in the electronegativity of the two atoms connected by a chemical bond. It is thus possible to assign a dipole moment to each bond in a complex molecule. The values of the dipole moments of the different molecules are tabulated:

Molecule	Dipole Moment/D	Comments
H_2	0	Symmetrical molecule, no displacement of charge.
H_2O	1.87	Lone pair of electrons on O, both —OH bonds have dipole which can be added vectorially.
Ar	0	Monatomic gas.
NH_3	1.47	Nitrogen atom has lone pair of electrons, each —NH bond has a dipole, which can be added vectorially.
CCl_4	0	Although each C—Cl bond has a considerable dipole, because the molecule is symmetrical (the four bond moments are directed tetrahedrally) their vector sum is zero
$CHCl_3$	1.02	Not a symmetrical molecule, hence residual dipole.
CO	0.12	Carbon is slightly less electronegative than oxygen.
CO_2	0	Although each C=O bond has an appreciable dipole, since the molecule is symmetrical there is no overall dipole.
C_6H_6	0	A symmetric molecule
$C_6H_6NO_2$	4.26	The nitro group is strongly electronegative, and so will be negatively charged with respect to the benzene ring.

****

SAQ 1.2a	You are considering changing the solvent in an electrochemical experiment from water to a series of methanol/water mixtures. If your electrolyte behaves as a strong electrolyte in water judge whether the following factors are important or not in making this solvent change.

(i) solute activity;   YES / NO
(ii) electrolyte solubility;   YES / NO
(iii) chemical nature of the electrolyte ions;
                                                                   YES / NO
(iv) ionic association.   YES / NO

Response

(i) You should have answered YES here. For a given electrolyte concentration, the activity is the product of the activity coefficient and the concentration. The activity coefficient can be calculated from eq Eq. 2i which includes the constant A. Values of A (Fig. 1.2f) are smaller for water than for methanol. The value of $y_{\pm}$ must therefore be larger in the water than methanol/water mixtures and the activity larger.

It is important to remember also that the ionic strength calculation assumes full dissociation of the electrolyte. This may not be a valid assumption for the electrolyte in the proposed solvent mixture.

(ii) Your answer should be YES. Methanol has a lower relative permittivity than water (Fig. 1.2a) and ions are consequently less stable in this solvent. This leads to a lower solubility in methanol/water than in water.

(iii) You should have answered NO here. All electrostatic effects of ions as discussed are based on their charge and not on their chemical nature. This chemical nature however, does have an effect as the concentration of the electrolyte increases. This effect is characteristic of each electrolyte and is not predictable.

(*iv*) Your answer should be YES. Ionic association increases with decreasing solvent relative permittivity and is thus likely to be significantly larger in methanol/water than in water. Whether this association will affect the electrochemical experiment will depend upon its extent. Separate experiments could be performed to establish the extent of the association.

****

SAQ 1.2b Estimate the mean molar ionic activity for the ions of lanthanum nitrate, $La(NO_3)_3$, in water at 25 °C in a 0.005 mol dm^{-3} solution.

Response

Your first reaction in reading the question should be that we cannot calculate the activity accurately since the concentration of the electrolyte is above the limit for the Debye Hückel limiting law. Nevertheless, the concentration is not too far above this limit for a sensible estimate to be made.

The first task is to calculate the ionic strength of the solution:

$$c(La^{3+}) = 0.005 \text{ mol dm}^{-3}$$
$$c(NO_3^-) = 0.015 \text{ mol dm}^{-3}$$

therefore,

$$I = 0.5 (0.005 \times 3^2 + 0.015 \times 1^2)$$
$$= 0.030 \text{ mol dm}^{-3}$$

Now we can estimate the mean molar ionic activity coefficient from the Debye Hückel limiting law,

$$-\lg y_\pm = A\,|\,z_+\,z_-\,|\,I^{\frac{1}{2}}$$
$$= 0.511 \times 3 \times 1 \times 0.030^{\frac{1}{2}}$$
$$= 0.265$$

therefore,

$$y_\pm = 0.543$$

Before the activity can be quoted we must determine the mean molar ionic concentration, (Eq. 1.2h),

$$n = 3$$
$$p = 1$$
$$c_\pm = 0.005 \times (1^1 \times 3^3)^{\frac{1}{4}}$$
$$= 0.0114 \text{ mol dm}^{-3}$$

The mean molar ionic activity is given by:

$$a_\pm = y_\pm\, c_\pm$$
$$a_\pm = 0.0543 \times 0.0114$$
$$= 0.00619$$

****

SAQ 1.2c　　Calculate the ionic strength of a mixture of 20 cm^3 of 0.10 mol dm^{-3} KNO$_3$ and 40 cm^3 of 0.20 mol dm^{-3} CaCl$_2$ which has been diluted to a final volume of 100 cm^3.

Response

Your answer should be 0.26 mol dm^{-3}.

The first task is to calculate the concentration of all ions in the final volume remembering that the potassium nitrate is diluted by a factor of 20/100 or 0.2 and the calcium chloride by a factor of 40/100 or 0.4.

The ionic concentrations are,

$$c(K^+) = 0.2 \times 0.10 = 0.02 \text{ mol dm}^{-3}$$
$$c(NO_3^-) = 0.2 \times 0.10 = 0.02 \text{ mol dm}^{-3}$$
$$c(Ca^{2+}) = 0.4 \times 0.20 = 0.08 \text{ mol dm}^{-3}$$
$$c(Cl^-) = 0.4 \times 0.40 = 0.16 \text{ mol dm}^{-3}$$

The ionic strength can now be deduced,

$$I = 0.5 \, (0.02 \times 1^2 + 0.02 \times 1^2 + 0.08 \times 2^2 + 0.16 \times 1^2)$$
$$= 0.5 \times 0.52$$

therefore,

$$I = 0.26 \text{ mol dm}^{-3}$$

****

SAQ 1.2d

> Calculate the ionic strength of a mixture of 20 cm^3 of 0.1 mol dm^{-3} sodium ethanoate and 10 cm^3 of 0.1 mol dm^{-3} hydrochloric acid made up to a total volume of 50 cm^3.

Response

The ionic strength of a solution is in Eq. 1.2c.

This definition is limited to completely dissociated electrolytes in solution. In this instance there are three equilibria present, those for the complete dissociation of sodium ethanoate and hydrochloric acid and one for the formation of undissociated (or partially dissociated) ethanoic acid:

$$NaOOCCH_3 \longrightarrow Na^+ + CH_3COO^-$$

$$HCl \longrightarrow H^+ + Cl^-$$

$$CH_3COOH \rightleftharpoons H^+ + CH_3COO^-$$

The partially dissociated ethanoic acid will not contribute to the total ionic strength. Thus hydrogen and ethanoate ions are removed from the original mixture forming ethanoic acid. Thus the ionic strength is effectively due to the sodium ethanoate (or its replacement NaCl) in solution. The concentration of the sodium and ethanoate ions are equal:

$$c_{Na+} = c_{CH_3COO^-} = 0.1 \times 20/50 = 0.04 \text{ mol dm}^{-3}$$

the factor 20/50 is included because there is a dilution factor. The value of z for Na^+ and CH_3COO^- is 1; thus the ionic strength is given by:

$$I = 0.5 (0.04 + 0.04) = 0.04 \text{ mol dm}^{-3}$$

It is important that you understand this type of calculation as it often is required to calculate the ionic strength of a buffer solution, which as you are aware consists of a weak acid and its conjugate base (or weak base and its conjugate acid).

****

SAQ 1.3a | In an electrochemical experiment, an aqueous electrolyte solution containing 0.1 mol dm^{-3} potassium chloride and 0.001 mol dm^{-3} zinc sulphate is the analyte solution. What percentage of the electrical current which passes through this solution will be carried by the zinc, potassium and chloride ions?

(Make use of the data in Fig. 1.3a. Neglect the effect of non-ideality on ionic mobility).

Response

The electrical current carried by any ion in solution is proportional to its molar ionic conductivity (λ). This parameter refers however to one mole of the ion in a volume of 1 m^3. At another concentration the value of λ must be multiplied by the ionic concentration (c). Eq. 1.3d allows the calculation of λ from the tabulated ionic mobility.

Summarising the above argument,

$$\text{current} \propto \text{conductance} \propto c\lambda \propto cuz$$

Applying to each solution ion,

$$I(K^+) \quad \propto 0.1 \times 7.62 \times 10^{-8} \times 1 \quad \propto 7.62 \times 10^{-9}$$
$$I(Cl^-) \quad \propto 0.1 \times 7.92 \times 10^{-8} \times 1 \quad \propto 7.92 \times 10^{-9}$$
$$I(Zn^{2+}) \quad \propto 0.001 \times 5.5 \times 10^{-8} \times 2 \quad \propto 0.11 \times 10^{-9}$$
$$I(SO_4^{2-}) \quad \propto 0.001 \times 8.25 \times 10^{-8} \times 2 \quad \propto 0.16 \times 10^{-9}$$

$$I(\text{total}) \quad \propto \qquad\qquad\qquad\qquad\qquad\qquad 15.81 \times 10^{-9}$$

From this table of values of the ionic currents we obtain, for the zinc ion the percentage current carried is:

$$\frac{0.11 \times 100}{15.81} = 0.70\%$$

for the potassium and chloride ions the percentage current carried is:

$$\frac{(7.62 + 7.92) \times 100}{15.81} = 98.3\%$$

You should note that solutions of the above composition are used in polarographic analysis where the zinc ion would be the ion for determination and the potassium chloride would be the 'swamping or supporting electrolyte'.

****

SAQ 1.4a   A cell is constructed from two copper wires dipping into aqueous copper(II)sulphate solution. What effects will be observed if:

(*i*) the two copper wires are joined together outside the cell,

(*ii*) a low voltage battery is connected between the two copper wires?

Response

You should first consider the chemical equilibrium at each copper electrode/solution interface. This will be,

$$Cu(s) \longrightarrow Cu^{2+}(aq) + 2e$$

An electrode potential will be established: let us assume that the metal is at a higher potential than the solution. We could depict the cell as:

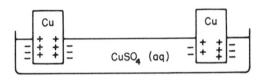

(*i*) Since the two copper electrodes are at the same electrical potential no current can flow between them when they are joined together.

(*ii*) If we now make one of the copper electrodes a higher potential than the other by connecting a battery between the two electrodes, electrons will be forced onto the low potential copper electrode. These electrons will remove copper ions from solution and copper deposition will occur. In like manner at the higher potential electrode, electrons will be produced by copper dissolving into the solution. Here we have formed an electrolytic or electrolysis cell.

Note that the electrons produced at the higher potential electrode will flow round the external circuit to the lower potential electrode thus ensuring an equal and opposite amount of reaction at the two electrodes.

****

SAQ 1.4b	An electric current was passed through a series of solutions of $AgNO_3$, $CrCl_3$, $ZnSO_4$ and $CuSO_4$. 1.000 g of silver was deposited from the first solution. Calculate (i) the quantity of electricity passed (ii) the current passing, and (iii) the weights of Cr, Zn and Cu deposited simultaneously [Ag = 107.9; Cr = 52.0; Zn = 65.4; Cu = 63.6].

Response

Faraday's laws can be summarized:

$$Q = It = F wz/A_r$$

where Q is the quantity of electricity passed/C; I the current/A, t the time/s; w the weight of substance deposited/g; z the valence of the element deposited,

A_r the relative atomic mass of the element deposited and F the faraday constant 96485 C mol^{-1}

(i)   Applying the equation for the deposition of silver enables us to calculate the total quantity of electricity passed:

$$Q = 96485 \times 1.000 \times 1/107.9 = 894.21C$$

(ii)   Remembering that $Q = It$ enables us to calculate the current passed:

$$I = Q/t = 894.21/60 \times 60 = 0.248A$$

(iii)   Using the same equation but rearranged to give w:

$$w = Q \times A_r/F \times z$$

for chromium w = 894.35 × 52.0/96485 × 3 = 0.161 g

zinc w = 894.35 × 65.4/96485 × 2 = 0.303 g

copper w = 894.35 × 63.6/96485 × 2 = 0.295 g

****

SAQ 2.1a

Predict the emf of a galvanic cell which has the following cell reaction ($z = 1$),

$$Ag\ (s) + 0.5\,Cl_2(g) \longrightarrow AgCl(s)$$

$$\Delta G^{\ominus} = -109.8\ kJ$$

Assume that the chlorine gas pressure is 1 atmosphere.

Response

Your answer should be 1.14 V.

You should substitute the values of $\Delta G^{\ominus}$, z and F into Eq. 2.1a remembering that $\Delta G^{\ominus}$ must be in J not kJ,

$$-109\ 800 = -1 \times E^{\ominus} \times 96485$$
$$E^{\ominus} = 1.14\ V$$

****

SAQ 2.2a   You have been given the task of devising an elec-
            trochemical cell for the estimation of chloride
            in a commercial product which has an aqueous
            base. Draw up a short list of half cells which
            you consider suitable for this purpose and then
            make a choice of two electrodes giving your rea-
            sons based on the criteria of:

            (*i*)   response;
            (*ii*)   practicability;
            (*iii*) availability;
            (*iv*)   safety.

Response

On your list you should have the following electrodes.

Indicator electrodes –   chlorine gas
                          silver/silver chloride
                          calomel

Reference electrodes –   silver/silver chloride
                          calomel
                          mercury(I)sulphate.

Chlorine gas electrode. This electrode is responsive to the chloride
and chlorine gas pressure. It would be necessary to keep the gas
pressure constant to make it responsive to the chloride ion only. It
is not a very practical electrode in view of the toxicity of the chlorine
and the need to prepare the platinised platinum electrode surface.
The electrode has to be made just prior to use and should only be
used if no other more suitable electrode is available.

Silver/silver chloride electrode. This electrode is responsive to the
chloride ion only. It is a very convenient laboratory electrode which
can be prepared easily and with which there are no safety problems.

As a reference electrode, there are occasions when it is not suitable because of the outflow of KCl from the liquid junction into the analyte solution.

Calomel electrode. The comments here are similar to those for the silver/silver chloride electrode. This electrode can be purchased commercially and is a very practical, reproducible electrode. There are, possibly, more problems with safety than the silver/silver chloride electrode because of the presence of mercury in the electrode.

Mercury(I)sulphate electrode. This reference electrode is often used where the diffusion of chloride ion from the electrode into the analyte solution is a serious problem. It is similar in construction to the calomel electrode and there are similar safety problems. The electrolyte within the electrode is potassium sulphate and this should be as dilute as possible, eg 0.1 mol dm^{-3}.

A reasonable choice of electrodes would be the silver/silver chloride indicator electrode with the mercury(I)sulphate as the references electrode.

****

SAQ 2.3a

A galvanic cell is constructed from two redox electrodes, one involving titanium(IV) and titanium(III) ions and the other involving cerium(IV) and cerium(III) ions. If the ions are all at unit activity and you are told that $E^{\ominus}(Ti^{4+}, Ti^{3+}) = -0.04$ V.

(i) Decide which electrode will be the positive electrode.

(ii) Write down the cell linear format. $\longrightarrow$

SAQ 2.3a   (*iii*) Write down the electrode and cell reac-
(cont.)   tions.

                    (*iv*) Calculate the free energy change for the
                           spontaneous cell reaction.

Response

Your answers should be:

(*i*)   Cerium redox electrode.

(*ii*) Pt | $Ti^{4+}(a = 1)$, $Ti^{3+}(a = 1)$ ‖ $Ce^{4+}(a = 1)$, $Ce^{3+}(a = 1)$ |
        Pt

(*iii*) LHE/*F* Ti^{3+} $\longrightarrow$ $Ti^{4+} + e$
        RHE/*F* $Ce^{4+} + e$ $\longrightarrow$ Ce^{3+}
        OCR/*F* $Ce^{4+} + Ti^{3+}$ $\rightleftharpoons$ $Ce^{3+} + Ti^{4+}$

(*iv*) $\Delta G^{\ominus} = -159$ kJ

In (*i*), we must consider the relative values of the standard electrode
potentials of the two redox electrodes. From Fig. 2.3b, the value for
the cerium couple is 1.61 V. Since this is more positive than the
titanium standard electrode potential, the cerium redox electrode
will form the positive electrode of the cell. The standard cell emf,

$$E^{\ominus}(\text{cell}) = E^{\ominus}(\text{RHE}) - E^{\ominus}(\text{LHE})$$
$$= 1.61 - (-0.04) = 1.65 \text{ V}.$$

(*ii*) The linear cell format has been given above. You should notice
that the activities of all the ions are quoted and the liquid–liquid
junction is indicated by the double vertical line.

In part (*iii*) of the question, oxidation must occur at the LHE and
therefore the titanium(III) ion is oxidised to titanium(IV) ion. This

is a one electron process as is the RHE reduction of cerium(IV) ion to cerium(III) ion. The overall cell reaction is obtained by adding these two electrode reactions. This is the spontaneous cell reaction.

The standard free energy change for the cell reaction is given by Eq. 2.1a,

$$\Delta G^{\ominus} = -zE^{\ominus}F = -1 \times 1.65 \times 96485$$
$$= -159200 \text{ J}$$
$$\therefore \quad \Delta G^{\ominus} = -159.2 \text{ kJ}$$

****

SAQ 2.4a

A galvanic cell is devised to monitor the cadmium ion activity in cadmium sulphate solutions at 25 °C. The cell consists of a cadmium indicator electrode and a saturated calomel reference electrode. Deduce for this cell:

(i)   the cell linear format;
(ii) the electrode and cell reactions;
(iii) the cell response equation;
(iv) the cell emf when the analyte is 0.1 mol dm^{-3} cadmium sulphate solution ($y\pm = 0.150$).

(use Figs. 2.2c, 2.3b, 2.4a as necessary)

Response

(i) Your answer should be:

$$Cd \mid CdSO_4(a) \parallel KCl(sat) \mid Hg_2Cl_2, \; Hg$$

From the tabulated data, the saturated calomel electrode potential at 25 °C is 0.241 V and the standard electrode potential for the cadmium electrode is −0.403 V. The calomel electrode must therefore be the RHE of the cell. This can be stated fairly confidently since these two potentials are quite different. If they were nearer together, the contribution of the analyte term in the Nernst equation could upset this prediction.

(ii) The cell reaction is, for 2 faradays:

$$Cd(s) + Hg_2Cl_2 \rightleftharpoons Cd^{2+} + 2Hg(l) + 2Cl^-$$

You can deduce this equation from the individual electrode reactions:

LHE/2F $\quad$ Cd $\quad\longrightarrow\quad$ $Cd^{2+} + 2e$
RHE/2F $\quad$ $Hg_2Cl_2 + 2e \longrightarrow 2Hg + 2Cl^-$
OCR/2F $\quad$ $Hg_2Cl_2 + Cd \rightleftharpoons Cd^{2+} + 2Hg + 2Cl^-$

(iii) The response equation is,

$$E = E(cal) - E^{\ominus}(Cd^{2+},Cd) + (2.303\,RT/2F)\lg\{1/a(Cd^{2+})\}$$

There are two ways to go about obtaining the cell response. Firstly we can look at the individual electrode responses, them add then together. In doing this we must remember that the LHE spontaneous reaction is oxidation but in using Eq. 2.3a it is assumed that the electrode reaction is a reduction. Thus:

RHE the response here is fixed at E(cal).

LHE the reaction is reduction,

$$Cd^{2+} + 2e \longrightarrow Cd$$

Thus,

$$E(Cd^{2+}, Cd) = E^{\ominus}(Cd^{2+}, Cd) + (2.303\,RT/2F)\lg a(Cd^{2+})$$

The cell response is then,

$$E = E(\text{RHE}) - E(\text{LHE})$$

$$E = E(\text{cal}) - E^{\ominus}(\text{Cd}^{2+}, \text{Cd}) - (2.303\,RT/2F)\,\lg a(\text{Cd}^{2+})$$

Alternatively we can take a second approach which is to apply the general Nernst Eq. 2.4a to the cell OCR remembering that the activity of the chloride ion is accounted for in the $E(\text{cal})$ term. Thus,

$$E = E(\text{cal}) - E^{\ominus}(\text{Cd}^{2+}, \text{Cd}) + (2.303\,RT/2F)\,\lg \{1/a(\text{Cd}^{2+})\}$$

(*iv*) Your answer should be 0.698 V

The activity of the cadmium ion is given by its concentration multiplied by its activity coefficient, ie

$$a(\text{Cd}^{2+}) = 0.1 \times 0.15 = 0.015$$

Putting this into the cell response equation, we have

$$E = 0.241 - (-0.403) + \frac{0.05915}{2}\,\lg\frac{1}{(0.015}$$
$$= 0.644 + 0.054$$

$$\therefore\quad E = 0.698\text{ V}$$

****

SAQ 2.5a   Calculate the ljp at 25 °C for the junction between each of the following pairs of electrolytes. Assume that activity coefficients are unity and that the transport number does not change with electrolyte concentration.

(*i*) $AgNO_3$ $c_1 = 0.005$ mol dm^{-3}: $c_2 = 0.01$ mol dm^{-3} $\longrightarrow$

SAQ 2.5a   (*ii*)   KCl $c_1 = 0.005$ mol dm^{-3}: $c_2 = 0.01$ mol
(cont.)   dm^{-3}

(*iii*) HCl $c_1 = 0.001$ mol dm^{-3}: $c_2 = 0.01$ mol
          dm^{-3}

Response

Your answers should be: (*i*) 1.28 mV (*ii*) 0.32 mV (*iii*) −11.4 mV

From Eq. 2.5a the term $2.303\,RT/F\,\lg\,(c_2/c_1)$ is common to all three electrolytes. Its value is,

$$59.15\,\lg\,(0.01/0.005) = 0.01781\ \text{V}$$

The difference in transport numbers $(t_- - t_+)$ can be evaluated from Fig. 1.3c For

(*i*) $(t_- - t_+) = 0.072$
∴ $E_\text{L} = 0.072 \times 0.01781 = 0.00128$ V $= 1.28$ mV

(*ii*) $(t_- - t_+) = 0.018$
∴ $E_\text{L} = 0.018 \times 0.01781 = 0.00032$ V $= 0.32$ mV

(*iii*) $(t_- - t_+) = -0.642$
∴ $E_\text{L} = -0.642 \times 0.01781 = -0.114$ V $= -11.4$ mV

You should note the very small value of E_L for the junction involving KCl where the cation and anion transport numbers are similar, and the large value of the ljp in HCl where the transport numbers are quite different. Note also the negative sign of this potential because the cation is the faster ion.

****

SAQ 2.6a	You are provided with a potentiometer having a galvanometer of sensitivity of 1 μA per mm deflection and a resistance of 100Ω. How accurately are you likely to be able to measure the emf of a cell whose internal resistance is 1900Ω? Is this an acceptable level of accuracy for 1% accuracy in concentration?

Response

Your answer should be that it should be possible to measure the emf to 2 mV. This is not accurate enough.

Look at Fig. 2.6b. The cell and galvanometer are connected as:

If a current flows through the galvanometer, a potential drop must occur from A to C so diminishing the effective cell emf. This potential drop = current × resistance.

If we assume that we can detect a 1 mm deflection in the galvanometer then this means that a current of 1 μA is our limit of detection. The total resistance is 1900 + 100 Ω and thus the potential drop is 10^{-6} × 2000 V or 2 mV.

You must then decide whether this is an acceptable level of accuracy or not. For $z = 1$, for a change in E of 1 mV the percentage change in concentration is approximately 4%. Thus for 2 mV, the value is 8%, a value far in excess of the required 1% accuracy.

****

SAQ 2.7a A pH meter shows the following pH readings at 25 °C for two standard buffer solutions of pH 4.00 and 8.90. The correct values are 4.00 and 9.18 respectively for these buffers. What mV output per pH unit is the meter receiving and what should it be receiving?

Response

The expected mV/pH unit is 59.15.

The observed mV/pH unit is 55.96.

The expected mV/pH unit is the Nernst factor at 25 °C; this is given in Fig. 2.4a as 59.15.

The electronic circuitry of the pH meter expects to receive this mV-value for every pH change in the cell solution. Since a pH difference of 8.90–4.00 = 4.90 units is registered the meter must only receive:

$$4.90 \times 59.15 = 289.8\text{mV}.$$

Since the true change in pH is 9.18–4.00 = 5.18 units the cell response is:

$$289.8/5.18 = 55.94\text{mV/pH unit}.$$

Such a loss in response would probably be due to the glass electrode. The above result could be quoted as a slope of 100(55.94/59.15) or 95%. The *slope* knob on the pH meter could compensate for this decrease in response.

****

SAQ 2.8a

> You have been asked to design an automatic titration system for the determination of the concentration of weak acids in aqueous solution. You have decided to use a cell consisting of glass and saturated calomel electrodes and an automatic titration procedure with sodium hydroxide as the titrant. The cell is known to show zero volts at pH = 7.
>
> Predict the cell emf at 25 °C during the titration of a weak acid ($K = 10^{-5}$) and approximate concentration 0.1 mol dm^{-3}, at:
>
> (*i*)   start of titration,
> (*ii*) half neutralisation.
>
> Assume that the activity coefficients of all species are unity, that there is zero ljp and that there is no volume change during the titration.
>
> (Clue: you must consider the acid dissociation equilibrium in a way similar to the use of the solubility product in the text above).

Response

Your answers should be (*i*) 237 mV, (*ii*) 118 mV.

Your first task is to decide on the response equation for the cell. This is given in Section 2.7 as,

$$E = E(\text{Cal}) - E'(\text{glass}) + 0.05915\text{pH}$$

Since at pH = 7, $E = 0$,

$$0 = E'(\text{glass}) - E(\text{cal}) - 0.05915 \times 7 \quad E'(\text{glass}) - E(\text{cal}) = 0.4141 \text{ V}$$

Thus, if we know the pH of any cell solution we can calculate the cell response. We must now study the acid dissociation equilibrium,

$$HA + H_2O \rightleftharpoons A^- + H_3O^+$$

$$K = \frac{c(A^-) \, c(H_3O^+)}{c(HA)}$$

If the acid concentration starts at c and the extent of dissociation is x then the equilibrium concentrations are

$$c(HA) = c(1 - x)$$
$$c(A^-) = cx$$
$$c(H_3O^+) = cx$$
$$\therefore \quad K = \frac{cx \, cx}{c(1 - x)}$$

The value of K is small, $pK = 5$ and $K = 10^{-5}$. Thus we can assume x is small at $c = 0.1 \text{ mol dm}^{-3}$.

We can replace $(1 - x)$ by unity, and the expression for K becomes:

$$K = cx^2$$

Having established this and remembering that the hydrogen ion concentration is given by cx, we can look at the cell response,

(*i*) at the start of the titration.

Here $x = (K/c)^{\frac{1}{2}}$

$$x = (10^{-5}/10^{-1})^{\frac{1}{2}} = 10^{-2}$$

$$\therefore \quad cx = 10^{-3}$$

$$\therefore \quad pH = 3$$

$$\therefore \quad E = 0.4141 - 0.05915 \times 3 = 0.2366 \text{ V} = 237 \text{ mV}$$

(*ii*) at half neutralisation. Here we know that the acid and its conjugate must have the same concentration, ie $c(\text{HA}) = c(\text{A}^-)$

$\therefore \quad K = c(\text{H}_3\text{O}^+) = 10^{-5}$

$\therefore \quad E = 0.4141 - 0.05915 \times 5 = 0.1183 \text{ V} = 118 \text{ mV}$

****

SAQ 3.1a

> Explain the differences in operation and terminology between an electrochemical cell and an electrolysis cell.

Response

Important differences include the following (3.1.1, 3.1.2)

(*i*) The electrodes participate chemically in the cell reaction in an electrochemical cell. In electrolysis the electrode is not changed during the cell reaction.

(*ii*) Electrochemical cells are operated under null current conditions, whereas in electrolysis a finite and usually changing current is passed continuously.

(*iii*) The spontaneous cell reaction in an electrochemical cell produces the cell emf. In an electrolysis cell an external voltage is applied to the cell, imposing a chemical reaction.

(*iv*) Negative (oxidation) and positive (reduction) electrodes in electrochemical cells are termed anodes (oxidation) and cathodes (reduction) in electrolysis cells.

(*v*) Due to polarisation effects if an increased current is required in an electrolysis cell then the external applied voltage must increase. For an electrochemical cell as current increases polarisation causes cell emf to fall.

****

SAQ 3.1b Define overpotential at an electrode. How will the potential of a reduction reaction occurring at a cathode change as the current in the cell is increased.

Response

The overpotential (3.1.2) η is defined:

$$\eta \;=\; \text{measured or applied} \;-\; \text{reversible or}$$
$$\text{electrode potential} \qquad \text{equilibrium electrode}$$
$$\text{potential}$$

Fig. 3.1f shows the general behaviour of electrode potential as the current increases in an electrolysis cell.

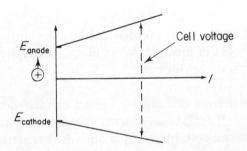

Fig. 3.1f.   *Electrode overpotential as a function of current*

Hence the answer is that the potential of the cathode will decrease. The overpotential will increase in a negative sense. The linear change in the E/I curve is only a crude approximation; an indication of the true shape is given in Fig. 3.1d.

****

| SAQ 3.1c | Give one example of a technique falling into each of the categories: microelectrolysis, macro-electrolysis, destructive, non-destructive. |

Response

There could be many answers but for example:

microelectrolysis : dc polarography
macroelectrolysis : controlled potential coulometry
destructive : atomic absorption spectroscopy
non-destructive : uv-visible absorption spectrophotometry.

The important point is that you can explain the four terms.

****

| SAQ 3.2a | Sketch and label a circuit for a potentiostat-ically controlled three electrode cell and explain how potentiostatic control is achieved and maintained. Why are three electrodes necessary? What problems arise when non-aqueous solutions are used? |

Response

A suitable circuit is that shown in Fig. 3.2c. Potentiostatic control is achieved by means of the auxiliary circuit in Fig. 3.2c. The reading on the high impedance voltmeter (P) has the significance of:

$$P = | E_{WE} - E_{RE} | = E_{WE} \text{ (SCE)}$$

if a saturated calomel electrode is used as a reference electrode. The required value of E_{WE} (SCE) is fed into the potentiostat memory and the potentiostat compares the required value with the measured value. Any difference is an error signal. The potentiostat causes the dc supply to alter in such a way (altering the cell current as a consequence) that the error signal is reduced to zero. When this is achieved (very rapidly) we have potentiostatic control. Control is maintained by the continuous comparing of required and measured signals and the consequent adjustment of the dc supply.

Three electrodes are necessary to overcome the fact that the potentials of most electrodes change when continuous current is passed (polarisation). In a two electrode cell the measured voltage is the difference in potential between two changing individual electrode potentials. In these circumstances it is impossible to calculate the potential of any one electrode.

Non-aqueous solvents cause problems in terms of solubility of analyte and supporting electrolyte but principally problems arise because the resistance of the solution increases. This causes loss of true potentiostatic control. The reference electrode can also become unstable.

****

SAQ 3.2b	Sketch a cell suitable for microelectrolysis and comment on the features. In what way would you change the design in order to carry out a typical macroelectrolysis.

Response

Your sketch should resemble Fig. 3.2g and the main features incorporated should be: (*i*) stirring, (*ii*) thermostatting, (*iii*) filling and emptying cell *in situ*, (*iv*) purging with gas, (*v*) gas venting, (*vi*) 3-electrodes, (*vii*) minimum volume.

In general for macroelectrolysis one would isolate the SE in a separate compartment in order to minimise contamination of the electrode reaction at the WE by products at the SE. One would also use a WE of much larger area. Later in your studies you will be able to compare controlled potential coulometry (4.2.2) an example of macroelectrolysis with dc polarography (3.5.1) an example of microelectrolysis.

****

SAQ 3.2c	Consider the following three electrolytes; (*i*) H_2O, (*ii*) 0.1 mol dm^{-3} aqueous KCl, (*iii*) 1.0 mol dm^{-3} $(C_2H_5)_4NBF$ in CH_3CN, with conductivities; (*i*) $\kappa = 10/^{-4}$ S m^{-1}, (*ii*) $\kappa = 1.289$ S m^{-1}, (*iii*) $\kappa = 5.55$ S m^{-1}. If two electrodes of equal area (0.2 cm^2) are placed in these solutions when a cell current of 20 μA is passing calculate the distance apart of the electrodes if an ohmic drop of <1 mV is to be achieved. Comment on the implications of your results in electro-analytical applications.

Response

G = conductance = κ/J, where κ is the conductivity and J the cell constant.

For our purposes we may assume that the area of the electrode is the geometric area, hence:

$$J = L/A = L/2 \times 10^{-5}$$

$$\therefore \quad G = \kappa/J = \kappa \times 2 \times 10^{-5}/L$$

Now $G = 1/R$, $\therefore R = L/\kappa \times 2 \times 10^{-5}$

By Ohm's law $V = IR$

$$\therefore \qquad\qquad V = \frac{20 \times 10^{-6} \times L}{\kappa \times 2 \times 10^{-5}} V$$

If $V = 10^{-3}$ V, then

$$L = \frac{10^{-3} \times \kappa \times 2 \times 10^{-5}}{20 \times 10^{-6}} \text{ m}$$

(i) $\kappa = 10^{-4}$ S m^{-1}, $\therefore L = 10^{-7}$ m

(ii) $\kappa = 1.289$ S m^{-1}, $\therefore L = 1.29 \times 10^{-3}$ m $= 1.29$ mm

(iii) $\kappa = 5.55$ S m^{-1}, $\therefore L = 5.55 \times 10^{-3}$ m $= 5.55$ mm

If you imagine using a Luggin probe as part of the auxiliary circuit in a three-electrode cell then it is clearly impossible to achieve an Ohmic drop of <1 mV for (i), pure H_2O. It is feasible to achieve an Ohmic drop of <1 mV for (ii) and (iii).

****

SAQ 3.3a	What are the five essential features of a good electrode/solvent/supporting electrolyte system?

Response

The five essential features are:

— a stable working electrode at which the required electrochemical reaction can take place efficiently;

— a stable secondary electrode capable of sustaining a satisfactory cell reaction;

— a homogeneous solution, ie all components are mutually soluble at all stages in the analytical procedure;

— a solution with a low electrical resistance;

— a voltage window with anodic and cathodic voltage limits as wide as possible.

You should be prepared to explain further what is meant by the required electrochemical reaction, and by voltage window and voltage limits.

****

SAQ 3.3b

What are the two roles of a supporting electrolyte? Calculate the relative magnitudes of the migration current and diffusion current for the zinc cation in an aqueous solution at 25 °C containing 5×10^{-4} mol dm^{-3} Zn(NO$_3$)$_2$ and 0.1 mol dm^{-3} KCl. Given mobility data (Fig. 1.3a):

Ion	$10^8 \ u/m^2 \ V^{-1} \ s^{-1}$
NO$_3^-$	7.4
Cl$^-$	7.9
K$^+$	7.6
Zn^{2+}	5.5

Response

The two roles are:

— to make the analyte solution sufficiently conducting by providing a low electrical resistance. This is particularly important when non-aqueous solvents are used.

— for those techniques, eg dc polarography, where diffusion of the analyte ions to the electrode is the required rate-determining mass transport process, the supporting electrolyte (present in great excess over the analyte) carries most of the migration current.

Answer to the calculation is $I_m/I_d = 0.0035$.

If you do not obtain this answer read 3.3.3 again and repeat the calculation before reading the rest of the response.

Ion	$z_i\, u_i\, c_i$	
Zn^{2+}	$2 \times 5.5 \times 10^{-8} \times 5 \times 10^{-4}$	$= 5.5 \times 10^{-11}$
K^+	$1 \times 7.6 \times 10^{-8} \times 0.1$	$= 7.6 \times 10^{-9}$
NO_3^-	$1 \times 7.4 \times 10^{-8} \times 10 \times 10^{-4}$	$= 7.4 \times 10^{-11}$
Cl^-	$1 \times 7.9 \times 10^{-8} \times 0.1$	$= 7.9 \times 10^{-9}$
	$\Sigma z_i\, u_i\, c_i$	1.56×10^8

$$t(Zn^{2+}) = \frac{z_i\, u_i\, c_i}{\Sigma\, z_i u_i c_i} = \frac{5.5 \times 10^{-11}}{1.56 \times 10^{-8}}$$
$$= 3.52 \times 10^{-3}$$

∴   A fraction 0.0035 of the total cathodic current is due to the migration of Zn^{2+} ions to the cathode.

ie $I_m \propto 0.035$ units, if total current is say 10 units.

∴ $I_d \propto 9.965$ units

$$\therefore \quad I_m/I_d = 0.035/9.965 = 0.0035$$

We see that under these conditions the transport of Zn^{2+} to the electrode is under diffusion control.

****

SAQ 3.3c	The standard reversible electrode potential for the reduction of hydronium ions in aqueous solution at 25 °C is -0.241 V (SCE) and yet hydrogen is evolved on the following surfaces at the potentials stated for a solution of 0.1 mol dm^{-3} aqueous HCl: Pt -0.321 V (SCE), Hg -1.363 V (SCE). Explain this and state how this relates to the choice of electrode material for a cathode. [2.303 $RT/F = 0.0592$ V at 25 °C]

Response

Your knowledge of basic electrochemistry (Part 2.0) will lead you to calculate first the actual reversible electrode potential in 0.1 mol dm^{-3} HCl. The answer is the same for both Pt and Hg.

$$E_e = E^\ominus - (2.303\,RT/F)\,\lg\{1/[H_3O^+]\}$$

This is the Nernst equation and yields:

$$E_e \quad -0.300V \text{ (SCE)}$$

The differences in the observed potentials (E) for hydrogen evolution on the two electrodes is due to the overpotential which is greater for Hg than it is for Pt. Advantage is taken of this in analytical methods where the analyte reaction occurs at a cathode. The cathodic voltage limit for Hg is much more negative than that for Pt.

****

SAQ 3.3d

The following information is available for gold as a possible electrode material. The reduction of hydronium ions in 1.0 mol dm^{-3} aqueous $HClO_4$ occurs at about -0.5 V (SCE). Gold is good as Pt on the anodic side provided complexing anions, eg CN^-, Cl^-, are absent. If these anions are present gold should not be used at potentials $> +0.5$ V (SCE).

Summarise your conclusions about Hg, C, Pt and Au as materials to be used in the construction of working electrodes.

Response

The required information relating to Hg, C and Pt is given in 3.3.1. You should come to the conclusion that:

Cathodic WE Hg $>$ C $>$ Au $>$ Pt

Anodic WE Pt $\simeq$ C $>$ Au $>$ Hg

This is based on the known voltage limits for these electrodes. Remember that carbon becomes more favoured when the analytes are organic in type.

****

SAQ 3.4a

List five techniques that belong in the category, voltammetry at finite current. What features do these techniques have in common?

Response

We have discussed ten such techniques: dc polarography, ac polarography, solid electrode voltammetry, amperometric titrations, linear sweep voltammetry, cyclic voltammetry, stripping voltammetry, sampled dc polarography, normal pulse polarography, differential pulse polarography. They have in common the fact that electrolysis is occurring, ie a finite current is passing through the cell directed by an external power source. The behaviour of a working electrode is monitored and the relationship between potential and current is used to determine the analyte concentration.

****

SAQ 3.4b	State the main contribution to the Faradaic current in an electrolysis cell. How would you arrange for diffusion control of the cell current and how does this type of control manifest itself?

Response

The Faradaic current is a measure of the rate of all the electrochemical reactions occurring at an electrode. If the Faradaic efficiency is 100% for a particular reaction, then only that reaction is occurring. The main contributions to the Faradaic current are:

(*i*) the rate of the overall electron transfer process occurring at the electrode surface, and

(*ii*) the rate of mass transport of the electroactive species through the solution to the electrode.

Diffusion control is arranged by the addition of a high concentration of a suitable supporting electrolyte to the solution.

Diffusion control manifests itself in the existence of a maximum
(limiting) current.

****

SAQ 3.4c   Draw the current/potential curves that you
           would expect for a reversible and irreversible re-
           action and explain the origin of the shapes of the
           curves.

Response

The shape of the curves depend upon the relative rates of the electron transfer process and the mass transport process.

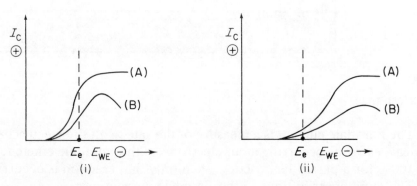

Fig. 3.4c. *I/E curves for reversible process (i) very fast
electron transfer and (ii) slow electron transfer*

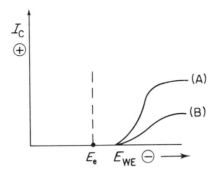

Fig. 3.4d. *I/E curves for irreversible process, very slow electron transfer*

In all these diagrams (A) represents fast mass transport and (B) represents slow mass transport.

The main feature of an irreversible process is that the potential at which the reaction is observed is displaced from E_e.

＊＊＊＊＊＊＊＊＊＊＊＊＊＊＊＊＊＊＊＊＊＊＊＊＊＊＊＊＊＊＊＊＊＊＊

SAQ 3.5a	Describe the principles of an electro-analytical method suitable for the determination of lead at the ppb level.

Response

The only method that we have discussed which is capable of determining heavy metals at the ppb level is differential pulse anodic stripping voltammetry (DPASV).

You should have described anodic stripping voltammetry (3.5.1g) and then discussed the use of differential pulse techniques in the stripping stage (3.5.1h). In describing the method you should have explained the need to purify the solvent/supporting electrolyte system in order to remove heavy metal impurities.

****

SAQ 3.5b What is the main factor that determines the limit of detection in dc polarography? How, in general, are the more advanced forms of polarography designed to overcome this factor?

Response

It is the capacitive current component of the total cell current which is the main factor in setting the detection limit in dc polarography. The more advanced forms of polarography all, in one way or another, discriminate between the Faradaic current and the capacitive current and in favour of the Faradaic current (3.5.1h).

****

SAQ 3.5c Describe a suitable electro-analytical method that could be used to determine nitrobenzene in acetonitrile solution at a concentration level of about 10^{-7} mol dm^{-3}.

Response

Here we have an organic analyte that requires a non-aqueous solvent, eg acetonitrile. You would need to use a quarternary ammonium salt as the supporting electrolyte and carbon might be preferred to mercury for the working electrode (cathode). Nitrobenzene is reduced to aminobenzene (aniline) the mechanism need not concern us here. To detect the analyte at the 10^{-7} mol dm^{-3} level a differential pulse method will be essential, eg differential pulse polarography (3.5.1h).

$$*********************************$$

SAQ 3.5d

The voltammograms for Br_2,Br^- and Sn^{4+},Sn^{2+} in aqueous solution at 25 °C obtained on a platinum working electrode are shown in Fig 3.5n.

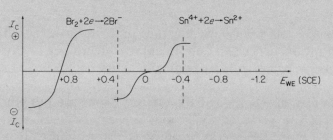

Fig. 3.5n. *Solid electrode voltammograms*

Draw the resulting amperometric titration result when a solution of tin(II) ions (Sn^{2+}) is titrated with bromine (Br_2) at a fixed working electrode potential of: (*i*) $E_{WE} = +0.3$ V (SCE), (*ii*) $E_{WE} = -0.4$ V (SCE)

Response

You should have predicted:

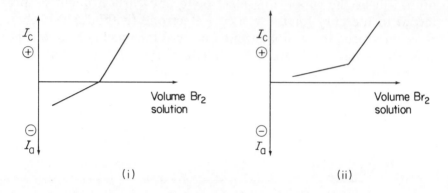

Fig. 3.5o. *Amperometric titration result*

(*i*) $E_{WE} = +0.3$ V (SCE); (*ii*) $E_{WE} = -0.4$ V (SCE)

In (*i*), at $+0.3$V (SCE), the Sn^{2+} ions are anodically active. As titration proceeds, anodic current decreases to the end-point. The excess bromine is cathodically active.

In (*ii*), at -0.4 V (SCE), the Sn^{2+} ions are not electroactive. However the product, Sn^{4+} ions are cathodically active and after the end-point the excess bromine is also cathodically active.

****

SAQ 4.1a Sketch out a classification of the methods of electro-analytical chemistry and correctly place the following techniques, cyclic voltammetry, chronopotentiometry, ion-selective electrode methods.

Response

The main sub-divisions are miscellaneous, voltammetry and coulometry.

Voltammetry is divided further into zero and finite current methods with the zero current methods known as potentiometry and the finite current methods usually termed just voltammetry.

Voltammetry at finite current is yet again divided into methods at controlled potential and methods at controlled current.

Coulometry is divided into methods at controlled potential and methods at controlled current. This classification can be displayed in various ways (4.1). Cyclic voltammetry should be placed under voltammetry at finite current (voltammetry) under controlled potential conditions. Chronopotentiometry belongs under voltammetry at finite current, controlled current conditions. Ion selective electrode methods belong to voltammetry at zero current (potentiometry).

****

SAQ 4.2a	Describe how you would remove heavy metal impurities from a 0.1 mol dm^{-3} aqueous solution of potassium chloride.

Response

The technique to use is controlled potential electrodeposition. This is a version of electrogravimetric analysis at controlled potential. The objective is not to weigh the amount of electroactive species deposited but just to remove them from solution. In all other respects, except one, the two techniques are identical, the exception is that a large mercury pool cathode is preferred to a platinum mesh. Set the potential of the working electrode about 10 mV more positive than the known potential at which potassium ions are reduced. This will cause all heavy metal cations to reduce since their reduction potentials are all more positive than that for the potassium cation. Details of the electrogravimetric analysis technique are given in 4.2.1.

****

SAQ 4.2b   An organic ketone may be determined quantitatively using bromine as a standard. Describe how you would determine such a ketone at the ppm level in aqueous solution using an electroanalytical method.

Response

You should have chosen the technique of coulometric titration. A bromine precursor is required, a soluble bromide. The bromide ions are oxidised at an anodic working electrode using a constant small current. A three-electrode galvanostatically controlled circuit is required. The bromine is generated in a controlled manner at a very low concentration which may be calculated using Faraday's law from the value of the current and the time of the experiment. The end-point may be determined in this case by visible spectrophotometry. The whole coulometry cell is placed in the spectrophotometer cell compartment and the wavelength is set on a strong absorbance wavelength for bromine. The titration is at the end-point when the absorbance starts to increase from an initial zero value.

****

SAQ 4.2c

> The phenol content of a river water sample is required. A 20 cm^3 sample is made slightly acidic and an excess of potassium bromide is added. It required a steady current of 8.6 mA for 187 s to produce the bromine required to quantitatively react with phenol by the reaction:
>
> $$PhOH + 3\,Br_2 \longrightarrow C_6H_2(OH)Br_3 + 3\,HBr.$$
>
> What is the concentration of phenol in the water in ppm assuming that the river water has a density of 1.00 g cm^{-3}?
>
> [$F = 96485$ C mol^{-1}; M_r(PhOH) = 94.1]

Response

Quantity of electricity $= 8.6 \times 10^{-3} \times 187$ C

$\qquad\qquad\qquad\qquad = 1.608$ C

$\qquad\qquad\qquad\qquad = 1.608/96485$ faraday

$\qquad\qquad\qquad\qquad = 1.666 \times 10^{-5}$ F

1 faraday produces 0.5 mol Br$_2$ (Br$^-$ + e $\longrightarrow$ 0.5 Br$_2$)

∴   1 Faraday will produce sufficient bromine to react with $\dfrac{1.666 \times 10^{-5}}{2 \times 3}$ mol phenol $= 2.777 \times 10^{-6}$ mol.

∴   20 cm^3 = 20 g water contains $2.777 \times 10^{-6} \times 94.1$ g phenol $= 2.613 \times 10^{-4}$ g phenol

∴   10^6 g water contains $\dfrac{2.613 \times 10^{-4} \times 10^6}{20}$ g phenol

$\qquad\qquad = 13$ g phenol

ie   13 ppm phenol.

Units of Measurement

For historic reasons a number of different units of measurement have evolved to express quantity of the same thing. In the 1960s, many international scientific bodies recommended the standardisation of names and symbols and the adoption universally of a coherent set of units—the SI units (Système Internationale d'Unités)—based on the definition of five basic units: metre (m); kilogram (kg); second (s); ampere (A); mole (mol); and candela (cd).

The earlier literature references and some of the older text books, naturally use the older units. Even now many practicing scientists have not adopted the SI unit as their working unit. It is therefore necessary to know of the older units and be able to interconvert with SI units.

In this series of texts SI units are used as standard practice. However in areas of activity where their use has not become general practice, eg biologically based laboratories, the earlier defined units are used. This is explained in the study guide to each unit.

Table 1 shows some symbols and abbreviations commonly used in analytical chemistry; Table 5 is a glossary of abbreviations used in this particular text. Table 2 shows some of the alternative methods for expressing the values of physical quantities and the relationship to the value in SI units.

More details and definition of other units may be found in the *Manual of Symbols and Terminology for Physicochemical Quantities and Units*, Whiffen, 1979, Pergamon Press.

Table 1   *Symbols and Abbreviations Commonly used in Analytical Chemistry*

Å	Angstrom
$A_r(X)$	relative atomic mass of X
A	ampere
E or U	energy
G	Gibbs free energy (function)
H	enthalpy
J	joule
K	kelvin (273.15 + t °C)
K	equilibrium constant (with subscripts p, c, therm etc.)
K_a, K_b	acid and base ionisation constants
$M_r(X)$	relative molecular mass of X
N	newton (SI unit of force)
P	total pressure
s	standard deviation
T	temperature/K
V	volume
V	volt (J A^{-1} s^{-1})
a, $a(A)$	activity, activity of A
c	concentration/ mol dm^{-3}
e	electron
g	gramme
I	current
s	second
t	temperature / °C
bp	boiling point
fp	freezing point
mp	melting point
≈	approximately equal to
<	less than
>	greater than
∝	proportional to
e, $\exp(x)$	exponential of x
$\ln x$	natural logarithm of x; $\ln x = 2.303 \log x$
$\lg x$	common logarithm of x to base 10

Table 2 *Alternative Methods of Expressing Various Physical Quantities*

1. **Mass (SI unit : kg)**

$$g = 10^{-3} \text{ kg}$$
$$mg = 10^{-3} \text{ g} = 10^{-6} \text{ kg}$$
$$\mu g = 10^{-6} \text{ g} = 10^{-9} \text{ kg}$$

2. **Length (SI unit : m)**

$$cm = 10^{-2} \text{ m}$$
$$\text{Å} = 10^{-10} \text{ m}$$
$$nm = 10^{-9} \text{ m} = 10\text{Å}$$
$$pm = 10^{-12} \text{ m} = 10^{-2} \text{ Å}$$

3. **Volume (SI unit : m³)**

$$l = dm^3 = 10^{-3} \text{ m}^3$$
$$ml = cm^3 = 10^{-6} \text{ m}^3$$
$$\mu l = 10^{-3} \text{ cm}^3$$

4. **Concentration (SI units : mol m^{-3})**

$$M = \text{mol } l^{-1} = \text{mol dm}^{-3} = 10^3 \text{ mol m}^{-3}$$
$$mg \, l^{-1} = \mu g \text{ cm}^{-3} = ppm = 10^{-3} \text{ g dm}^{-3}$$
$$\mu g \, g^{-1} = ppm = 10^{-6} \text{ g g}^{-1}$$
$$ng \text{ cm}^{-3} = 10^{-6} \text{ g dm}^{-3}$$
$$ng \text{ dm}^{-3} = pg \text{ cm}^{-3}$$
$$pg \, g^{-1} = ppb = 10^{-12} \text{ g g}^{-1}$$
$$mg\% = 10^{-2} \text{ g dm}^{-3}$$
$$\mu g\% = 10^{-5} \text{ g dm}^{-3}$$

5. **Pressure (SI unit : N m^{-2} = kg m^{-1} s^{-2})**

$$Pa = Nm^{-2}$$
$$atmos = 101\,325 \text{ N m}^{-2}$$
$$bar = 10^5 \text{ N m}^{-2}$$
$$torr = mmHg = 133.322 \text{ N m}^{-2}$$

6. **Energy (SI unit : J = kg m^2 s^{-2})**

$$cal = 4.184 \text{ J}$$
$$erg = 10^{-7} \text{ J}$$
$$eV = 1.602 \times 10^{-19} \text{ J}$$

Table 3   *Prefixes for SI Units*

Fraction	Prefix	Symbol
10^{-1}	deci	d
10^{-2}	centi	c
10^{-3}	milli	m
10^{-6}	micro	μ
10^{-9}	nano	n
10^{-12}	pico	p
10^{-15}	femto	f
10^{-18}	atto	a

Multiple	Prefix	Symbol
10	deka	da
10^2	hecto	h
10^3	kilo	k
10^6	mega	M
10^9	giga	G
10^{12}	tera	T
10^{15}	peta	P
10^{18}	exa	E

Table 4 *Recommended Values of Physical Constants*

Physical constant	Symbol	Value
acceleration due to gravity	g	9.81 m s^{-2}
Avogadro constant	N_A	$6.022\,05 \times 10^{23} \text{ mol}^{-1}$
Boltzmann constant	k	$1.380\,66 \times 10^{-23} \text{ J K}^{-1}$
charge to mass ratio	e/m	$1.758\,796 \times 10^{11} \text{ C kg}^{-1}$
electronic charge	e	$1.602\,19 \times 10^{-19} \text{ C}$
Faraday constant	F	$9.648\,46 \times 10^{4} \text{ C mol}^{-1}$
gas constant	R	$8.314 \text{ J K}^{-1} \text{ mol}^{-1}$
'ice-point' temperature	T_{ice}	$273.150 \text{ K exactly}$
molar volume of ideal gas (stp)	V_m	$2.241\,38 \times 10^{-2} \text{ m}^3 \text{ mol}^{-1}$
permittivity of a vacuum	ϵ_0	$8.854\,188 \times 10^{-12} \text{ kg}^{-1} \text{ m}^{-3} \text{ s}^4 \text{ A}^2 \text{ (F m}^{-1})$
Planck constant	h	$6.626\,2 \times 10^{-34} \text{ J s}$
standard atmosphere pressure	p	$101\,325 \text{ N m}^{-2} \text{ exactly}$
atomic mass unit	m_u	$1.660\,566 \times 10^{-27} \text{ kg}$
speed of light in a vacuum	c	$2.997\,925 \times 10^{8} \text{ m s}^{-1}$

Table 5 *Glossary and Abbreviations used in Electrochemistry*

A	area; Debye–Hückel constant
C	coulomb
D	ionic diffusion coefficient
D	Debye
E	emf
$E(X^+, X)$	electrode potential of X^+, X
$E_{0.5}$	half-wave potential
E_L	liquid junction potential
F	Faraday constant
G	Gibbs (free) energy function, conductance
I	ionic strength, current
i	current density
J	cell constant
L	length
Q	electric charge, quantity of electricity
R	resistance
S	siemens
V	potential difference
e	electron, electronic charge
$f_\pm$	mean ionic activity coefficient (mole fraction basis)
n	number of electrons transferred
t	time
$t(X)$	transport number of X
$u(X)$	ionic number of X

Table 5 *Glossary and Abbreviations used in Electrochemistry (Continued)*

$y_\pm$ mean ionic activity coefficient (molarity basis)

z charge number of ion

$\gamma_\pm$ mean ionic activity coefficient (molarity basis)

η overpotential

κ conductivity

μ dipole moment

$\lambda(X)$ ionic molar conductivity of X

Ω ohm

Superscripts
$\ominus$ indicating standard value of a property

∞ indicating value of a property at infinite dilution

Subscripts
i referring to typical ionic species

$+,-$ referring to positive, negative ion

Other Abbreviations

ASV	Anodic Stripping Voltammetry
DME	Dropping Mercury Electrode
DMF	dimethylformamide
DMSO	dimethylsulphoxide
DPASV	Differential Pulse Anodic Stripping Voltammn
HIV	High Impedance Voltmeter
HMDE	Hanging Mercury Drop Electrode
LHE, RHE	Left (Right) Hand Electrode
lpj	Liquid Junction Potential
MFE	Mercury Film Electrode
OCR	Overall Cell Reaction
RE	Reference Electrode
SCE	Saturated Calomel Electrode
SE	Secondary Electrode
SHE	Standard Hydrogen Electrode
TISAB	Total Ionic Strength Adjustment Buffer
WE	Working Electrode